SPECTROSCOPIC TECHNIQUES FOR
ORGANIC CHEMISTS

SPECTROSCOPIC TECHNIQUES FOR
ORGANIC CHEMISTS

James W. Cooper
Tufts University

A WILEY-INTERSCIENCE PUBLICATION

JOHN WILEY & SONS
New York   Chichester   Brisbane   Toronto

Library of Congress Cataloging in Publication Data

Cooper, James William, 1943–
  Spectroscopic techniques for organic chemists.

  "A Wiley-Interscience publication."
  Includes index.
  1. Spectrum analysis.   2. Chemistry, Organic.
I. Title.

QD272.S6C66   547'.308'5   79-23952
ISBN 0-471-05166-7

Printed in the United States of America

10 9 8 7 6 5 4 3 2

PREFACE

Since spectroscopy has become the main tool by which even beginning organic chemists identify compounds and study their interactions, this book has been designed to help advanced undergraduates and beginning graduate students learn these important techniques. A large number of illustrations have been included in each chapter so that students can learn in two ways, through the visual and the numerical aspects of spectroscopy, and thus reinforce their knowledge. Also included are a number of problems with solutions, so that students can test their understanding of the material as they study. Since the solutions are rather detailed, students can not only check their answers but also check their reasoning, step by step. Instructors wishing additional problems, without solutions, may consult the numerous problem books referred to at the end of each chapter.

In addition to the usual topics of infrared, proton nmr, and mass spectroscopy, this text includes a chapter on the use of the increasingly routine technique of ^{13}C nmr spectroscopy and a chapter discussing a nonmathematical treatment of the Fourier transform techniques used in modern nmr. Furthermore, we have found it useful to precede the treatment of UV spectroscopy with a brief introduction to simple Hückel MO theory, which gives additional insight into the meaning of electronic energy levels. Since timesharing computers are now an accepted part of modern college life, two timesharing versions of computer programs are included: one is for the calculation of simple Hückel energy levels, rather than deriving the hand-calculation method in detail; the other is the LAOCOON III program for the iteration of theoretical nmr spectra for a best fit with experimental spectra.

JAMES W. COOPER

Medford, Massachusetts
February 1980

v

ACKNOWLEDGMENTS

In undertaking a book containing hundreds of spectra, one naturally asks many favors and obtains the spectra from many sources. All of the infrared spectra were obtained on a Perkin–Elmer 727B Infrared Spectrometer running in slow scan (resolution) mode, giving a nominal resolution of 3 cm^{-1}. The water vapor and CO_2 spectrum in Figure 1.1 was provided through the cooperation of Professor Peter Griffiths of Ohio University. The spectrometer block diagram in Figure 1.4 is courtesy of Perkin–Elmer.

Almost all the proton nmr spectra were obtained from the Aldrich Library of Nmr Spectra through the courtesy of Dr. Charles Pouchert and the Aldrich Chemical Company. Many of the remaining spectra were obtained on the Tufts University Chemistry Department EM-360A and HA-100. The decoupled spectra of crotonaldehyde in Figure 4.24 were obtained through the courtesy of Bruker Instruments, Inc., of Billerica, Massachusetts. The high field spectra in Figure 4.29 were obtained by Dr. Joseph Dadok of Carnegie-Mellon University on their new 600 MHz spectrometer in correlation mode.

Most of the ^{13}C nmr spectra are plotted using a PDP-11 spectrum generation and plotting program of my own devising from data listed in the Bruker ^{13}C Data Bank, Volume 1. The remainder are actual spectra obtained on a WP-80 spectrometer at Bruker Instruments, Inc.

The mass spectra were obtained from the catalog compiled by Stenhagen et al., noted in Reference 1 of Chapter 9, and plotted using a PDP-11 plotting program of my devising on a high-precision plotter loaned to us by Dr. Ben Perlman of the Tufts Mechanical Engineering Department. Actual mass spectra were run by William Cote and Mike Santorsa of this department.

The Raman spectra in Chapter 10 were obtained by Dr. Robert Mooney and Mary Ann Hazle of the Sohio Research Center in Cleveland, Ohio.

I would also like to acknowledge the enthusiastic teaching of Professor Robert Ouellette with whom I first studied organic spectroscopy and from whom I have borrowed one or two examples, and of Professor Gideon Fraenkel who first made SHMO theory clear to me and showed me how you find the wavelength of a tennis ball. I also want to thank Professor Aksel Bothner-By for permission to reproduce the listing of LAOCOON III.

Finally, I would like to thank the numerous people who have read and commented on the manuscript: Dr. Kenner Christensen of Northwestern University, Drs. Terry Haas and Michael Delaney of Tufts University, Dr. Felix Wehrli and James Schroer of Bruker Instruments, Inc., and Chemistry 155 students Charles Thibault, Robert Kerstein, James O'Brien, George Pawle, Peter Teng, and Karel Janac.

The manuscript was typed most expeditiously by Natalie Camelo and Laurie Lydon, and the diagrams were provided by Robert Sheetz.

J. W. C.

CONTENTS

THREE

Introduction to Nuclear Magnetic Resonance   **53**

FOUR

Chemical Interpretation of Proton Nuclear Magnetic Resonance Spectra   64

SIX

^{13}C Nuclear Magnetic Resonance Spectroscopy   165

SEVEN

Simple Hückel MO Theory   201

EIGHT

Ultraviolet Spectroscopy

NINE

Introduction to Mass Spectroscopy

TEN

Raman Spectroscopy   **301**

SPECTROSCOPIC TECHNIQUES FOR
ORGANIC CHEMISTS

CHAPTER ONE
INTRODUCTION TO
INFRARED SPECTROSCOPY

IR spectroscopy is the fastest and cheapest of the spectroscopic techniques used by organic chemists. It is simply the measurement of the absorption of IR frequencies by organic compounds placed in the path of the beam of light. The samples can be solids, liquids, or gases and can be measured in solution or as neat liquids mulled with KBr or mineral oil. Thus spectra can be obtained in just a few minutes from partially purified materials in order to give an indication that the reactions have proceeded as desired.

THE IR REGION OF THE ELECTROMAGNETIC SPECTRUM

While it is common in physics to discuss light in terms of its wavelength, this is only one of two common ways of referring to IR absorptions. Let us represent the actual wavelength by λ, and recall that

$$v' = \frac{c}{\lambda} \tag{1.1}$$

where v' is the frequency of the light of wavelength λ and c is the speed of light. (We are reserving v, unprimed, to represent wave *numbers* as we will see below.) We can refer to an IR wavelength as a number of micrometers (or microns)

$$\lambda = 6.2 \times 10^{-6} \text{ cm} = 6.2 \ \mu\text{m}$$

or as a frequency. We could calculate the frequency from Eq. 1.1 and find that

$$v' = \frac{2.99 \times 10^8 \text{ m/sec}}{6.2 \times 10^{-6} \text{ m}} = 4.82 \times 10^{13} \text{ sec}^{-1}.$$

This frequency is a large number and rather difficult to deal with, so that IR spectroscopists, recognizing that wavelength and frequency are reciprocally

1

related, have selected a proportionality constant such that the frequencies are numbers of a convenient size with which to work. Thus they have defined the quantity *wave number* or *v* as

$$v = \frac{1}{(\lambda \text{ in cm})}$$

or

$$v = \frac{10^4}{(\lambda \text{ in } \mu\text{m})} \qquad (1.2)$$

and we then find that our wavelength of 6.20 μm becomes

$$= \frac{10^4}{(6.20 \ \mu\text{m})} = 1613 \text{ wave numbers.}$$

The units of wave numbers are expressed as cm^{-1} or in kaysers. We thus refer to the units of wave numbers as "centimeters to the minus 1" or more commonly as "reciprocal centimeters." Note that wave numbers are directly proportional to frequency and thus to the energy of the absorption while wavelengths are inversely proportional to energy, since

$$E = hv'.$$

Wave numbers are not strictly frequencies, but are proportional to frequency, and are thus commonly referred to as if they were frequencies. The actual relationship between v and v' is the speed of light c. If we express λ in cm,

$$v = \frac{v'}{c} = \frac{1}{\lambda}. \qquad (1.3)$$

What are the frequencies or wavelengths which make up the electromagnetic spectrum? We divide light into infrared, visible, and ultraviolet in order of increasing energy, and place microwaves below and X-rays above. To compare them we will look at their relative wavelengths:

	Wavelength	
Name	Meters	Common Units
Microwaves	$1 \times 10^{-1} - 1 \times 10^{-6}$	1 mm–10 cm
Infrared	$1 \times 10^{-4} - 8 \times 10^{-7}$	100–0.8 μm
Visible	$8 \times 10^{-7} - 4 \times 10^{-7}$	800–400 nm
Ultraviolet	$4 \times 10^{-7} - 100 \times 10^{-8}$	400–100 nm
X-Rays	$1 \times 10^{-8} - 5 \times 10^{-11}$	100–0.5 Å

In this chapter we will be concerned only with IR frequencies, of which only the middle of the range is of interest to organic chemists. We divide the infrared region into the near, mid, and far IR as follows:

	Near IR	Mid IR	Far IR
Frequency	14,300–4000 cm^{-1}	4000–650 cm^{-1}	650–200 cm^{-1}
Wavelength	0.7–2.5 um	2.5–15 um	15–100 um
Phenomena	overtones of	vibrations	absorptions of ligands
	C—H absorptions	and bending	and other low-energy
			species

ABSORPTIONS OF ORGANIC MOLECULES

The reason that most organic chemists find IR spectroscopy of great interest in their work is that most carbon–hydrogen, carbon–carbon, and carbon–oxygen bonds stretch at frequencies in the mid-IR region. Furthermore,

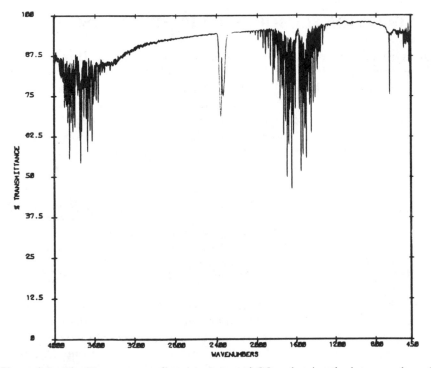

Figure 1.1 The IR spectrum of water vapor and CO_2, showing the large number of lines due to vibrational transitions. Courtesy of Professor Peter Griffiths, Ohio University.

these stretchings, rockings, and other motions are characteristic of the type of compound and of its functional groups, resonance possibilities, and shape so that the chemist can rapidly identify a number of important structural features from an IR spectrum.

If we look at a high resolution gas phase IR spectrum of some simple molecule,[1] as shown in Figure 1.1, we see that there are a veritable forest of lines stemming from the various vibrational energy levels and the smaller rotational energy levels which lie between them. These complex spectra are not amenable to simple analysis, but fortunately they are not the type of spectra with which we generally have to deal. In the liquid phase, the rotational energy levels simply broaden out the vibrational transitions, leading to a number of simpler, broad lines. These lines are characteristic of various functional groups, substitution patterns, and π-overlap as we will see below. While the lines are usually unsplit or *singlets*, we refer to related closely spaced lines as *doublets*, *triplets*, and so forth.

VIBRATIONAL MODES IN IR SPECTROSCOPY

Obviously, there are many possible vibrations in a molecule. However, only those stretchings which cause a change in dipole moment will show an IR absorption. Those which show no change in dipole moment may be observed by Raman spectroscopy (Chapter 10) and are often of less interest to organic chemists. To describe the types of vibrations, let us consider the simple water molecule. Since it is a bent molecule, we would expect to see various wagging and scissoring motions as common stretches. These are illustrated below:

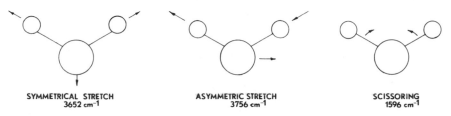

SYMMETRICAL STRETCH
3652 cm^{-1}
ASYMMETRIC STRETCH
3756 cm^{-1}
SCISSORING
1596 cm^{-1}

Note that the asymmetric stretch is of somewhat higher energy than the symmetric stretch, and that both require much more energy than scissoring vibrations. These are classical stretching motions and occur exactly as drawn.

We now extend these vibrational modes to a —CH_2— group, assuming that the group is anchored so that only the C and the two H's are actually in motion. Clearly other groups attached to the CH_2 affect the nature of the stretching which actually occurs.

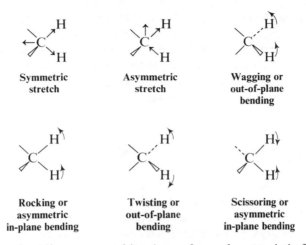

| Symmetric stretch | Asymmetric stretch | Wagging or out-of-plane bending |

| Rocking or asymmetric in-plane bending | Twisting or out-of-plane bending | Scissoring or asymmetric in-plane bending |

Molecules also vibrate at combinations of two characteristic frequencies such as $v_1 + v_2$ and $v_1 - v_2$. These combination bands are not always clearly identified since they may be weak and/or in regions of the spectrum containing many other lines.

One interesting phenomenon which is occasionally observed occurs when a fundamental frequency is near an overtone from some other frequency. The result is an increase in the intensity of the overtone and a decrease in the fundamental. In the spectrum of cyclopentanone[2] shown in Figure 1.2, we

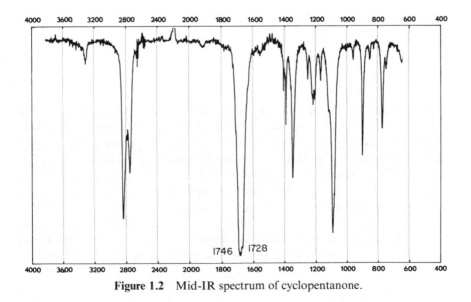

1746 1728

Figure 1.2 Mid-IR spectrum of cyclopentanone.

see a doublet with maxima at 1728 and 1746 cm^{-1}. This frequency is generally a carbonyl-stretching frequency range, and there is clearly only one kind of carbonyl in cyclopentanone. However, there is a ring-breathing frequency at 889 cm^{-1} (which is observed only in the Raman spectrum) whose overtone interacts with the carbonyl frequency, and perturbs it slightly. This phenomenon is known as *Fermi resonance*.

BOND STRETCHINGS IN IR

As a first approximation, we can regard this stretching as similar to that occurring when a spring with weights of varying mass on each end is stretched. The mass effects of having a larger and larger " weight " on one end have been tabulated[3] for the C—H stretch in a series of tetramethyl compounds where the central atom M varies as shown in Table 1.1.

Table 1.1   Mass Effects on the Stretch Frequency in M(CH$_3$)$_4$

M	C—H Symmetric bend
C	1370
Si	1254
Ge	1244
Sn	1205
Pb	1169

It is clear that the larger M is, the lower the frequency of the stretch. This fact can also be used to pick out various frequencies by substituting a deuterium for a hydrogen and looking for the shifted frequency in the IR spectrum. For example, Figure 1.3 shows the IR spectra of CHCl$_3$ and CDCl$_3$. If true C—H stretches are being examined, the ratio of the frequencies should be about 1.414, since the vibration of a simple harmonic oscillator depends on the square root of the reduced mass. In these spectra, the two higher frequency lines have the same ratio, $3035/2250 = 1224/910 = 1.34$, while the lowest lines do not have this ratio. The upper two lines are due to the C—H (C—D) stretching and rocking, and the lowest line is due to a C—Cl stretch.

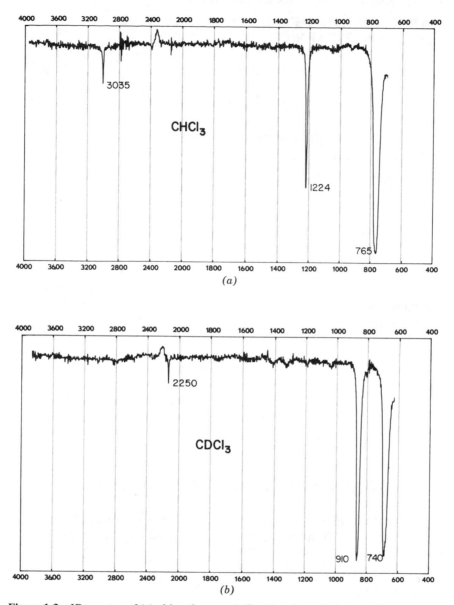

Figure 1.3 IR spectra of (*a*) chloroform and (*b*) chloroform-*d*, showing the variation in stretching frequency with bond strength.

ABSORPTION OF LIGHT IN OPTICAL SPECTROMETERS

Light absorption is governed by Beer's law:

$$A = \log \frac{1}{T} = \log \frac{I_0}{I} = Kcb \qquad (1.3)$$

where I = transmitted radiation intensity
I_0 = intensity of incident radiation
A = absorbance
T = transmittance
c = concentration
b = pathlength
K = a proportionality constant for a given compound

This logarithmic relationship can be valuable in quantitative work, but in most routine IR measurements the spectra are plotted as wave number versus percentage of transmittance ($\%T$), which is not linearly related to concentration.

If we express b in cm and c in moles/l then we can write

$$A = \varepsilon cb \qquad (1.4)$$

where ε is a molar proportionality constant called *molar absorptivity*. We will use this value in UV spectroscopy in Chapter 8, but will usually refer to the intensities of lines in IR spectra only as strong, medium, or weak (s, m, or w).

COMPONENTS OF IR SPECTROMETERS

An IR spectrometer can be considered to consist of a number of functional blocks: the source of radiation, the sample, a grating, the detector, and the output, usually a recorder.

Source of Radiation

The most common source of IR radiation for the IR spectrometer is a coil of Nichrome wire with a ceramic core, which glows when an electric current is passed through it. This source is relatively cheap and reliable but less intense than some other types found in research grade spectrometers. The Globar, a more intense source, is often found in Fourier transform IR spectrometers. It consists of a silicon carbide rod about 50 mm long and 4 mm in diameter which has a voltage applied across it. Its resistance increases with time, and some facility must be provided for increasing the voltage if a Globar source is used.

The Grating

The IR light source provides a broad spectrum of light which must, of course, be dispersed into monochromatic components. Any device which does this is known as a *monochromator*. Older spectrometers used NaCl prisms to perform this task, but modern spectrometers use a diffraction grating. The grating consists of a number of equally spaced grooves ruled onto a metal surface and spaced such that light of only one wavelength (and its multiples) will undergo constructive interference at a given angle from the grating. Light of different wavelengths can be obtained by rotating the grating relative to the light source, thus scanning through various frequencies.

The Detector

Most detectors are simply thermocouples whose emf varies with the light (heat) incident upon them. One type consists of a piece of blackened gold foil welded to two dissimilar metals chosen to give a large thermoelectric emf.

A more sensitive but less common detector is the Golay detector, which consists of a sealed cell of xenon gas having a blackened metal plate for incident light to fall on and a silvered diaphragm at the rear. As the plate is warmed by the light, it expands the trapped gas which causes distortion of the diaphragm. A light source striking the back of the diaphragm is then detected by phototube, amplifying this distortion.

Still other detectors include those based on the photoconductivity of semiconductors and those based on resistance thermometers.

SPECTROMETER DESIGN

The major functions of the spectrometer are shown in the block diagram in Figure 1.4. The light is generated at the source and reflected by two mirrors to two light paths: one through the sample and one through a reference cell. After passing through the sample and reference chambers, the light beams are recombined using a sector mirror that alternates one set of optics with the other so that the light from the two beams alternates in the light path about 10–25 times a second. The beam is then collimated using a pair of slits and is reflected from the diffraction′ grating through another slit to the detector.

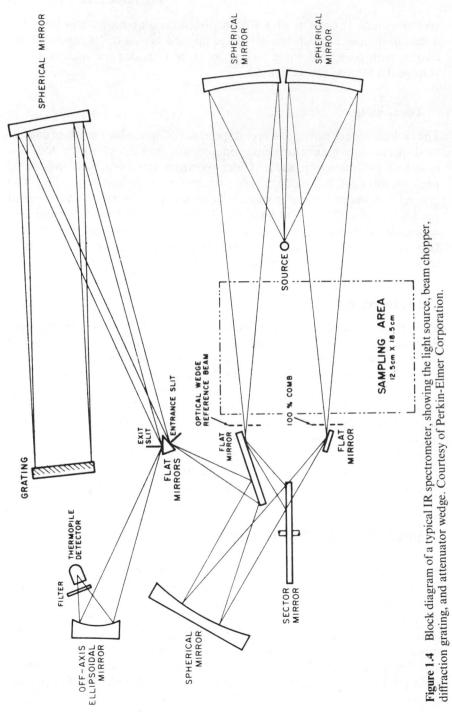

Figure 1.4 Block diagram of a typical IR spectrometer, showing the light source, beam chopper, diffraction grating, and attenuator wedge. Courtesy of Perkin-Elmer Corporation.

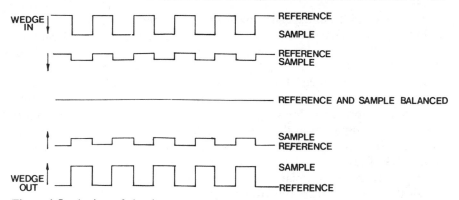

Figure 1.5 Action of the detector on the attenuator wedge to null the square wave generated by differences in intensity from the sample and reference beams.

The frequency of the light arriving at the detector is selected by the rotation of the grating and other extraneous frequencies may be filtered out by inserting filters into the light path at various ranges in the scan.

The light arriving at the detector is uniform in intensity over time as long as no sample is placed in the sample chamber. When a sample is inserted its absorption will decrease the intensity in the beam when the chopper has selected the sample beam. Thus the light arriving at the detector will be a square wave with a high intensity when the reference beam comes through and a low intensity when the sample beam comes through. The detector responds to this square wave by trying to null it. This is accomplished by attenuating the reference beam by lowering the variable density wedge into the beam until it is of the same intensity as the sample beam. When the light frequency striking the detector reaches a region where there is less difference between the sample and the reference beam, the wedge is removed until the square wave is again nulled.

Clearly the motion of this attenuator wedge parallels the intensity of the sample's IR absorption at a given frequency. Thus the pen on the chart recorder follows the motion of the wedge, moving down when the wedge moves down and up when the wedge moves up. This is illustrated in Figure 1.5.

SAMPLE PREPARATION FOR IR SPECTROSCOPY

Liquid Samples

The simplest way to obtain IR spectra of liquid samples is to place a drop or two on a salt plate and press another against it in a sample holder. If the sample is extremely volatile or has a fairly low absorbance in the region of interest, a small Teflon spacer ring may be inserted between the two salt disks to increase the volume of the sample making up the liquid film. Sodium chloride disks are used most commonly, but cell windows are also made of KBr, CaF_2, CsBr, CsI, ZnSe, AgCl, and AgBr. In addition, windows are available that are made of KRS-5, which is a mixed thallium bromide–iodide compound, and Irtran-2, which is made by hot pressing powdered ZnS.

Sodium chloride windows are by far the cheapest and are for the most part totally transparent in the region of interest to organic chemists. However, they are opaque below 650 cm^{-1}, which may be a disadvantage for some work. They are also brittle and, of course, quite soluble in water and to some extent in other polar compounds. Thus NaCl windows must be polished or replaced from time to time, but polishing is fairly easily accomplished. Cells are normally stored in a desiccator when they are not in use. All of the spectra in this book were obtained using NaCl windows, and thus cut off below 650 cm^{-1}.

The cell holder most favored for routine liquid film spectra is the simple one shown in Figure 1.6, which is simply a press fit O-ring supported holder that holds the two plates in contact fairly accurately. Windows are usually cleaned in CCl_4 since it is quite volatile and has few absorptions if a trace remains. Water, of course, is to be scrupulously avoided.

Figure 1.6 A simple press-lock IR cell holder, showing an NaCl disk.

Solid Samples as Pellets

Solid samples can be handled most accurately by carefully mixing a few milligrams of the solid with some carefully dried spectroscopic grade KBr in an agate or mullite mortar or in a Wig-L-Bug electric agitator. This resulting mixture is placed in a stainless steel die and, for most precise results, evacuated using a vacuum pump. The resulting sample is then pressed in a hydraulic press at a pressure of about 30 tons/in.^{2} (4500 kg/cm^2) for a few minutes. The piston of the press and the base of the press are then removed, and the barrel can be placed directly in the IR sample holder.

In addition to the obvious trouble of this technique, it is not always easy to get a clear pellet that is sufficiently transparent. Furthermore, these pellets are fragile and may break when the piston and base are removed.

To simplify the production of pellets for student use, several manufacturers have produced minipresses which have a smaller diameter and allow the student to apply pressure by simply tightening one polished bolt-end against another. The bolts are then unscrewed and a pellet remains in the middle. Obviously as much pressure cannot be applied with two wrenches on the bolt as can be applied in a hydraulic press, but for suitably chosen samples it is possible to make usable pellets. However, it should be noted that since the diameter of these minipresses is smaller, the total transmitted light is smaller and resolution suffers. Figure 1.7 shows examples of high-grade pellet dies and a minipress for student use.

Since the pellet that is produced may on occasion be relatively opaque, the attenuation of the sample beam by the spectrometer may be so great that the plot of the spectrum will run off the bottom of the chart paper, especially at the high frequency end. A reference beam attenuator, a sort of comb placed in the reference beam, can be used to lower the intensity of the reference beam and bring the baseline back to near the top of the chart (100% T). Such attenuators usually reduce resolution somewhat.

Liquid Mulls

A simpler way of obtaining spectra of solid samples is by mulling a small amount of sample with a liquid such as mineral oil (Nujol). The resulting spectrum will be clear except for the regions around 2900 and 1400 cm^{-1}. However, since the 2900 cm^{-1} region is present in nearly all organic compounds from the C—H symmetric and asymmetric stretch, this information is seldom needed. The spectrum of mineral oil (Nujol) is shown in Figure 1.8.

Other liquids can be used as mulls, of course, if these regions are of interest. The most common mulling liquids are Fluorolube, a chlorofluorohydrocarbon which is transparent at 2900 cm^{-1} but opaque below 1650 cm^{-1},

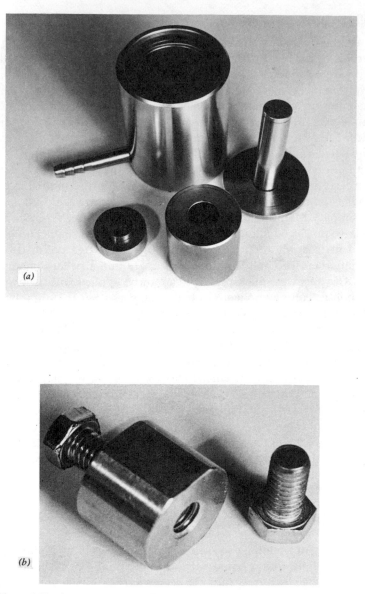

Figure 1.7 A vacuum KBr pellet press and a simple two-bolt minipress.

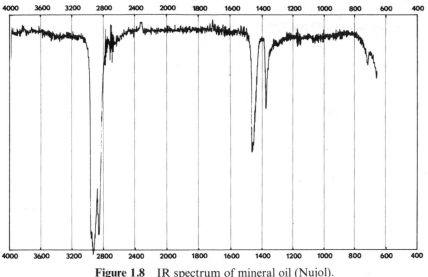

Figure 1.8   IR spectrum of mineral oil (Nujol).

and hexachlorobutadiene which is transparent above 1650 cm^{-1} and between 1500 and 1250 cm^{-1}. Comparison of spectra obtained by mulling samples in several liquids will allow examination of the entire mid-IR.

Solution IR

High-quality IR spectra of solids or liquids can be obtained in solution. This necessitates careful choice of a solvent that not only will dissolve the compounds of interest but will also be transparent in most of the mid-IR region. Unfortunately, there is no one solvent that is transparent in the entire region, and the solvent to be used must be selected with regard to the lines one expects to see. The most popular solvents are CS_2 and CCl_4, since CS_2 has bands only at 1400–1600 cm^{-1} and 2100–2200 cm^{-1} and CCl_4 has bands only at 700–850 cm^{-1}. However, even these bands can be minimized by using two *matched* solution cells: one in the sample beam and one in the reference beam. In this case, the peaks due to the solvent should, in theory, be subtracted out. While the solvent peaks may not appear, any small peaks near the solvent absorption may also be subtracted out and only large ones will be visible. Furthermore, any peaks lying under the solvent bands will be lost completely since no energy will be reaching the detector from either the sample or the reference beam. This is illustrated in Figure 1.9 where the spectrum of ethylbenzene is shown in CCl_4 solution with and without a matched solution cell and as a liquid film without any solvent. Note that with the matched cell, the peak at 1000 cm^{-1} becomes visible.

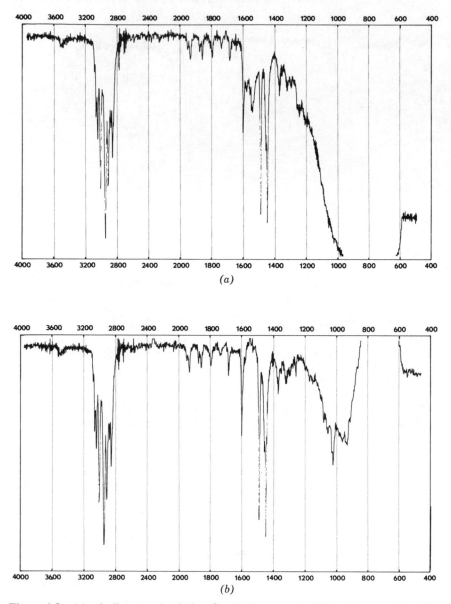

Figure 1.9 (*a*) ethylbenzene in CCl_4, (*b*) ethylbenzene in CCl_4 with matching CCl_4 solution cell in reference beam, and (*c*) neat ethylbenzene liquid film.

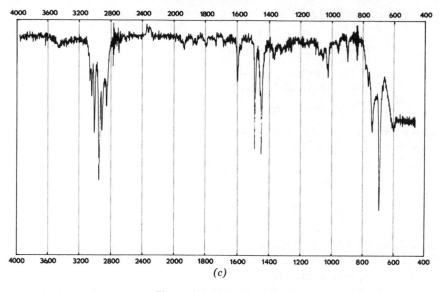

(*c*)

Figure 1.9   (*Continued*)

Table 1.2 lists some of the more common solvents used in IR spectroscopy. Clearly, spectra must be run in several solvents to obtain a spectrum of the entire mid-IR region.

Table 1.2   Opaque Regions in Common IR Solvents

Solvent	Opaque Regions (cm^{-1})	
CCl_4	700–850	
CS_2	1400–1600	2100–2200
$CHCl_3$	600–820	1175–1250
C_2Cl_4	750–950	
Benzene	600–750	3000–3100
CH_2Cl_2	600–820	1200–1300
Acetone	1100–1850	2800–3000
Cyclohexane	2600–3000	
Ethyl ether	1050–1200	2700–3000
Hexane, heptane	1400–1500	2800–3000
DMSO	900–1100	
Toluene	600–750	2800–3200

PRESENTATION OF THE IR SPECTRA

We divide IR spectra into three regions

The functional group region	$4000-1300$ cm^{-1}
The fingerprint region	$1300-910$ cm^{-1}
The aromatic region	$910-650$ cm^{-1}

Peaks in the functional group region are characteristic of particular functional groups, such as —OH, CH, —NH, —SH, C=O, C=C—H, and C≡C—H, as well as some overtones. Peaks in the fingerprint region are a very complex mixture of overtones and rocking frequencies which are not usually assignable. However, the fingerprint region of any IR spectrum is unique and thus is a "fingerprint" of the compound that is unlikely to be found even in other quite similar compounds. A strong peak in the aromatic region means that the compound is aromatic, although there are some exceptions; however, no compound can be considered aromatic *unless* it does have one or more strong peaks in this region.

Examining a typical IR spectrum such as that in Figure 1.9, we find that the scale of the spectrum is not entirely linear. While most modern IR spectrometers produce plots that are linear in wave number, there is a point at 2000 cm^{-1} where the scale is doubled. Thus, peaks from 2000–4000 cm^{-1} occupy less than half of the plotted chart, while peaks from 600–2000 cm^{-1}, a smaller range of frequencies, occupy a much greater distance along the chart. This is done in most spectrometers so that the greater detail of the fingerprint region can be observed more closely. This often produces a small positive or negative "blip" where the gear change occurs. Older prism spectrometers produce IR spectra that are linear in wave*length* rather than wave number, often with a break in the resolution around 5 μm (2000 cm^{-1}). In both cases, spectra are presented with 100% transmittance at the top and 100% absorption at the bottom, so that the "peaks" are all pointing downwards.

CALIBRATION OF THE IR SPECTRA

Since there is some variability in the positioning (and indeed the printing) of IR chart paper, the exact positions of the peaks may be called into question even when precisely ruled chart paper is used. The usual calibration standard is polystyrene film, whose spectrum is shown in Figure 1.10. The peak at 1603 cm^{-1} is a particularly sharp one which is commonly recorded on IR spectra as a reference if exact calibration is desired. If there are peaks in this region, the line at 1946 cm^{-1} is a good second choice. Rather than using

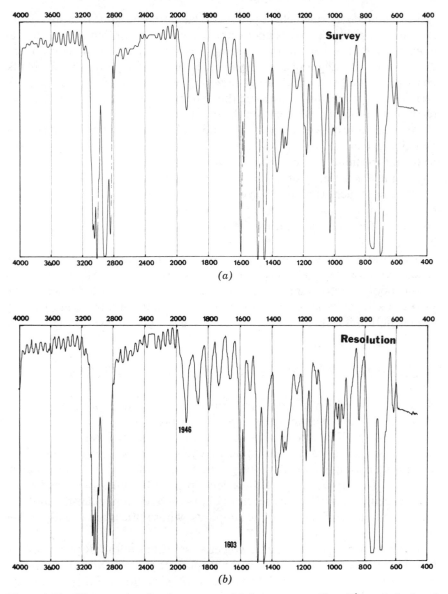

Figure 1.10 IR spectrum of polystyrene at (*a*) fast scan rate (6 cm^{-1} resolution) and (*b*) slow scan rate (3 cm^{-1} resolution).

expensive precisely ruled chart paper, we have found that it is more advantageous to use blank chart paper or paper with only an outline of the frequency scale and then overlay a semitransparent piece of precisely ruled chart paper. By aligning the polystyrene calibration peaks on the blank chart paper with the correct frequency on the calibrated paper, exact peak assignments can be made without the expense of high-quality chart paper or the confusion of offset lines on the chart itself.

An even more convenient method is possible with the advent of office copy machines which can copy IR spectra onto transparencies. To use this method, a copying machine must be found that does not reduce the chart slightly while copying it. Then a copy of the accurately printed chart is made onto a clear plastic sheet, such as is used for overhead projection transparencies. Each spectrum to be calibrated is then placed under the transparent sheet with the 1603 cm^{-1} line aligned, and the frequencies can be read off from the chart lines.

It should be noted that many routine IR spectrometers have two scanning speeds: a survey speed and a resolution or slow scan speed. All of the spectra in this book were obtained at the slower speed. At the faster speed, sharp peaks blur together and pen skipping becomes more of a problem. Usually the resolution is only about half as good. This is also shown in Figure 1.10.

USE OF PENS WITH IR SPECTROMETERS

No discussion of any scientific instrument that produces plotted output is complete without mentioning the perennial problem of keeping the pens writing. Most IR spectrometers come supplied with several small felt-tip pens, which are adequate for rapid analysis of a few spectra, but which become increasingly difficult to keep writing, and whose felt points broaden with use. The replacement of these felt-tip pens becomes more and more of an expense since they cost several dollars each, even in quantity, and can only be purchased from scientific suppliers, whose prices are usually rather high.

We have found that there are two better solutions to the IR pen problem. One is the use of Leroy or Kohinoor Rapidograph pens with those instruments designed to hold them. These pens will write forever if they are kept clean and filled with ink regularly. They can be kept from drying out by storing them in a jar with a small amount of water, and they can be kept from clogging by shaking them several times to make the cleaning wire dislodge any dried ink particles in the tip. If the pen still does not write, refill its ink supply and while the ink reservoir is off, unscrew the tip and rinse it under running water. *Never* squeeze the plastic ink reservoir to make the pen write.

Then reassemble the tip and ink reservoir and dry it. It should start writing with only one or two shakes. Any type of ink can, of course, be used in these pens, but if India-type drawing ink is used we have found Pelikan to be the smoothest and least likely to clog.

The most desirable pen for student use is one of the newer smooth-writing ballpoint pens adapted to fit the pen holder of the spectrometer. These pens are inexpensive, are purchased easily, and can be fastened into a holder using a grommet. The best choice is a ballpoint refill since they are shorter and thus wobble less in the holder.

REFERENCES

1. P. R. Griffiths, *Chemical Infrared Fourier Transform Spectroscopy*, Wiley-Interscience, New York, 1975.
2. C. L. Angel *et al.*, *Spectrochim. Acta*, 926 (1959).
3. C. W. Young, J. S. Koehler, and D. S. McKinney, *J. Am. Chem. Soc.*, **69**, 1410 (1947).

CHAPTER TWO
INTERPRETATION OF
INFRARED SPECTRA

The infrared spectra of organic compounds will always be dominated by the spectra of the carbon chains themselves. Accordingly, we take up the spectra of hydrocarbons first and then examine the spectra of various substituent groups.

SPECTRA OF HYDROCARBONS

Alkanes

The IR spectra of hydrocarbons consist of C—H and C—C stretching and bending frequencies (see Table 2.1). The symmetric and asymmetric stretch frequencies are all in the 2900s and all overlap in most spectra. The bending frequencies are around 1400 cm^{-1} and also overlap much of the time. In addition, there is a set of methylene twisting and wagging frequencies between 1150 and 1350 cm^{-1} which are all fairly weak. Note that the asymmetric stretch frequencies are always at higher frequencies since this deformation puts the atoms closer together than usual.

Functional groups having chains shorter than four carbons have slightly higher rocking frequencies, which can often be diagnostic:

$$734\text{--}743 \text{ cm}^{-1} \text{ propyl chain}$$

$$770\text{--}790 \text{ cm}^{-1} \text{ ethyl chain}$$

There are also some special spectral details which we can often recognize as due to branching of chains. The most common of these is the *gem*-dimethyl doublet. Whenever two methyl groups are attached to the same carbon, the interaction between in-phase and out-of-phase symmetrical CH$_3$ bending leads to a doublet: two lines of approximately equal intensity between 1370 and 1385 cm^{-1}. It is important to note that the *tert*-butyl group contains *gem*-dimethyls as well and the doublet can also be observed in compounds containing it. However, the *tert*-butyl doublet is less symmetric, with the lower line usually having a greater intensity than the upper one.

Table 2.1 Hydrocarbon Stretching and Bending Frequencies

Group	Stretching Frequencies (cm⁻¹)	Bending Frequencies (cm⁻¹)
—CH_3	2962 asymmetric stretch 2872 symmetric stretch	1450 asymmetric bend 1375 symmetric bend
—CH_2—	2926 asymmetric stretch 2853 symmetric stretch	1465 scissor
		720 rocking (for chains > 4 carbons)

Tertiary hydrogens in pure hydrocarbons can often be recognized by the fact that the fingerprint region usually consists of a number of bands, while *n*-alkanes have almost no lines in this region at all.

With these simple rules in mind, let us look at the spectrum in Figure 2.1. The compound has the molecular formula C_6H_{14} and has only a few important lines in it. First, we recognize from the formula that the compound must be an alkane and that the broad lines around 2900 cm⁻¹ must be due to the C—H symmetric and asymmetric stretching frequencies. We look around 1370–1380 cm⁻¹ and find a doublet due to bending frequencies; we conclude that the compound contains a *gem*-dimethyl group. We now look around the

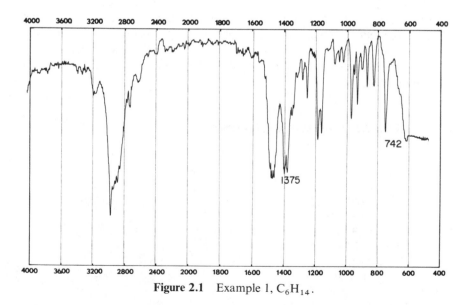

Figure 2.1 Example 1, C_6H_{14}.

700 cm^{-1} region for further information and find that there is a line at 742 cm^{-1}, which is characteristic of a propyl group. The only compound we can draw having *gem*-dimethyls and a propyl group is

$$\underset{\quad}{CH_3-CH_2-CH_2-\overset{\overset{\displaystyle CH_3}{|}}{CH}-CH_3}$$

2-methylpentane. For confirmation, we recognize that this compound contains a tertiary hydrogen, and we examine the fingerprint region for multiple bands and we do indeed find them.

Alkenes

Hydrogens attached to alkene carbons are held somewhat more tightly to that carbon (the bonds have more *s*-character) and thus stretch at a slightly higher frequency

$=CH_2$ 3080 cm^{-1} asymmetric stretch

2975 cm^{-1} symmetric stretch (low intensity)

$=CHR$ 3020 cm^{-1} asymmetric stretch

Because the stronger C—H stretching frequencies are above 3000 cm^{-1}, we usually refer to 3000 cm^{-1} as the line of *unsaturation*: any compound having C—H stretches above 3000 cm^{-1} contains some unsaturation.

The carbon–carbon stretching frequencies in alkenes are also at a higher energy (frequency), as we would expect. We usually look for a weak to medium band in the mid-1600 region. We can deduce from the intensity and frequency a little information regarding the substitution of the double bond as shown in Table 2.2.

Table 2.2 C=C Stretching Frequencies

Group	Stretching Frequency (cm^{-1})	Intensity
RCH=CH$_2$	1645	medium
R$_2$C=CH$_2$	1655	
cis-RCH=CHR	1660	
trans-RCH=CHR	1675	
tri and tetra substituted	1670	weak

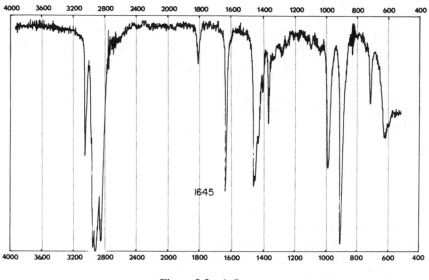

Figure 2.2   1-Octene.

Figure 2.2 shows a spectrum of 1-octene, showing the C=C stretch at 1645 cm^{-1}.

Cycloalkanes

Cycloalkanes and cycloalkenes have no particularly unique bands compared to ordinary alkanes, with the exception of cyclopropanes which are so strained that their stretching frequency is above the "line of unsaturation."[1,2]

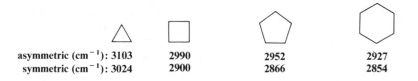

asymmetric (cm^{-1}): 3103	2990	2952	2927
symmetric (cm^{-1}): 3024	2900	2866	2854

Epoxides are an exception, absorbing in the range of 2990–3080 cm^{-1}, but it is seldom that you have an epoxide-containing compound without advance knowledge. They also show characteristic bands at 1250 cm^{-1} and 950–810 cm^{-1} due to the symmetric and asymmetric stretching of the ring.

Alkynes

In keeping with our rule regarding the line of unsaturation, alkynes, whose hydrogens are held even more tightly than alkenes, absorb at 3270–3315 cm^{-1}. Since the C—C bonds have more p-character, the C—H bond has more s-character and the CH stretch requires more energy. Note, however, that this C—H stretch will be present only in terminal alkynes and that interior alkynes will thus show no lines above 3000 cm^{-1}. The C≡C stretch in alkenes is fairly weak, but can usually be recognized:

internal C≡C   2210–2260 cm^{-1}

external C≡C   2100–2140 cm^{-1}

Symmetrically substituted alkynes will show no C≡C stretching band at all, since there is no dipole moment change for this stretch. Figure 2.3 shows the spectrum of 1-hexyne, showing the highly unsaturated C—H stretch at 3305 cm^{-1} and the terminal C≡C stretch at 2110 cm^{-1}. The only other triple bond commonly found in organic chemistry is the nitrile group C≡N, which has a strong absorption at 2260–2210 cm^{-1}. It is more intense than the alkyne stretch because there is a much greater change in the dipole moment in stretching the C—N bond.

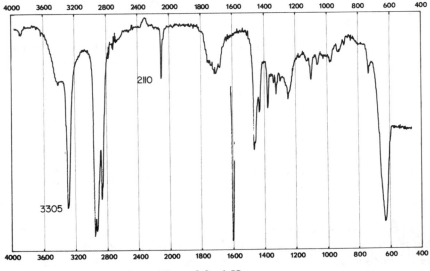

Figure 2.3 1-Hexyne.

AROMATIC HYDROCARBONS

Aromatic hydrocarbons have large numbers of C=C and C—H stretching and bending vibrations. The most prominent is the strong C—H out-of-plane bend between 900 and 675 cm^{-1}. There must be at least one *strong* absorption in this region if the compound is to be considered aromatic. In addition, there is a C=C stretch at 1575–1600 cm^{-1}, which is often a doublet. The C—H asymmetric and symmetric stretch occurs in the 3000–3100 cm^{-1} range and thus agrees with our rule about the line of unsaturation. The spectrum of toluene is shown in Figure 2.4.

Substituted benzene rings give weak but extremely diagnostic patterns[3] between 1650 and 2000 cm^{-1}. These patterns can be described as follows:

Monosubstituted	Four distinct "fingers," the strongest at 2000 cm^{-1}
Ortho-substituted	Two triplets
Meta-substituted	Two lines and a triplet at the low frequency end
Para-substituted	Two lines, the stronger one at higher frequency

These patterns are illustrated in Figure 2.5. The patterns are seldom as strong as shown in the figure, where the spectra were obtained by using a Teflon spacer to increase the amount of liquid in the light path. Note that they are much weaker in Figure 2.4.

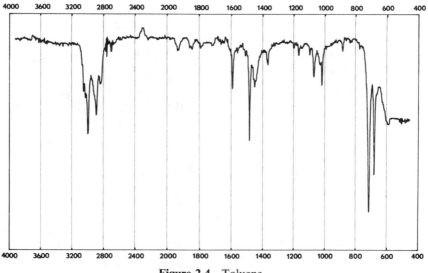

Figure 2.4 Toluene.

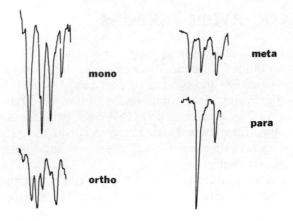

Figure 2.5 Characteristic patterns observed in the region 1650–2000 cm^{-1} for mono-, *ortho*-, *meta*-, and *para*-substituted benzenes.

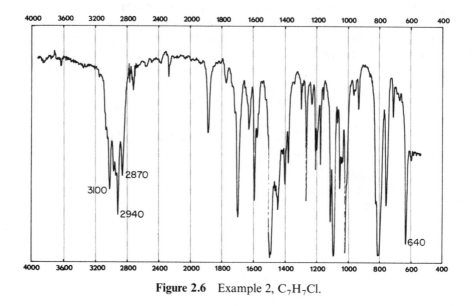

Figure 2.6 Example 2, C$_7$H$_7$Cl.

Additional confirming data may be obtained by examining the C—H out-of-plane bend frequencies:

Monosubstituted	770–730 cm^{-1} and 710–690 cm^{-1}
Ortho-substituted	770–735 cm^{-1}
Meta-substituted	810–750 cm^{-1} and 710–690 cm^{-1}
Para-substituted	860–800 cm^{-1}

Let us consider the spectrum in Figure 2.6 having the formula C_7H_7Cl. The strong peak near 800 cm^{-1} clearly indicates its aromaticity. The line at 3100 cm^{-1} also indicates unsaturation. However, the lines at 2870 and 2940 cm^{-1} indicate the presence of a saturated C—H stretch and a toluene is thus indicated. The simple pattern between 1650 and 2000 cm^{-1} indicates the *para*-substitution pattern, confirmed by the 820 cm^{-1} line, and the compound is thus *para*-chlorotoluene.

CARBONYL COMPOUNDS

Perhaps the most distinctive absorption in IR spectra is that of the carbonyl functional group. The carbonyl always gives a strong, sharp line in the range of 1580–1900 cm^{-1}, but with most ketones, aldehydes, and acids generally occurs in the mid 1700s. Carbonyl stretching frequencies are affected by electrical effects which withdraw or release electrons from the carbon–oxygen double bond. For example, we see a great increase in the frequency (energy) when more and more fluorines are added as substituents to the carbons attached to the carbonyl:[4,5]

CH_3—C—CH_3	CF_3—C—CH_3	CF_3—C—CF_3	F—C—F
‖	‖	‖	‖
O	O	O	O
1724 cm^{-1}	1769 cm^{-1}	1801 cm^{-1}	1928 cm^{-1}

The spectrum of acetone is shown in Figure 2.7. Note the strong, sharp carbonyl absorption at 1725 cm^{-1}.

Electrical effects through an adjacent aromatic ring have also been documented for *para*-substituted acetophenones:[6]

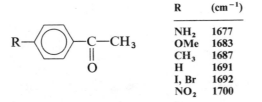

R	(cm^{-1})
NH_2	1677
OMe	1683
CH_3	1687
H	1691
I, Br	1692
NO_2	1700

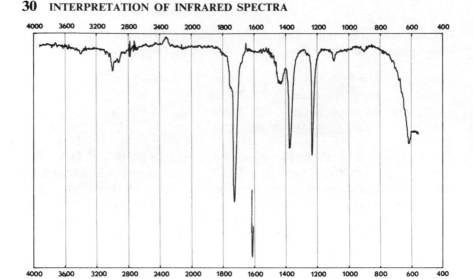

Figure 2.7 Acetone.

Note that the more electron releasing the group, the lower the carbonyl stretching frequency; the more withdrawing, the higher the frequency. This is readily rationalized by realizing that when the carbonyl stretches there is some charge redistribution so that the carbon becomes more positively charged and the oxygen more negatively charged:

$$C{=}O \quad \xrightarrow{h\nu} \quad \overset{\delta+ \quad \delta-}{C{\cdots}O}$$

Conjugation also markedly decreases the carbonyl absorption frequency, since this also will stabilize the small charge separation developed as the group stretches. Thus we find that α,β-unsaturated ketones absorb in the region of 1660–1680 cm^{-1}. Similarly, aromatic ketones are also stabilized and absorb below 1700 cm^{-1}. We can, in many cases, make the generalization that a carbonyl absorption below 1700 cm^{-1} must be in conjugation and refer to this as another line of unsaturation.

On the other extreme, cyclic ketones tend to increase in absorption frequency as the strain of the ring becomes greater:[7]

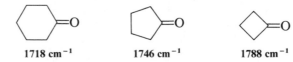

1718 cm^{-1} 1746 cm^{-1} 1788 cm^{-1}

We explain this by realizing that as the ring contracts, there is more p-character in the C—C bonds and thus more s-character to the C=O bond, requiring more energy to stretch it.

ALDEHYDES

Aldehydes generally absorb about 10–15 cm^{-1} above the corresponding ketone and show the same sort of electrical and conjugative effects as do ketones. We also find two weak C—H stretching peaks in the region of 2700–2900 cm^{-1}. One of these is often exactly at 2720 cm^{-1} and thus is considered

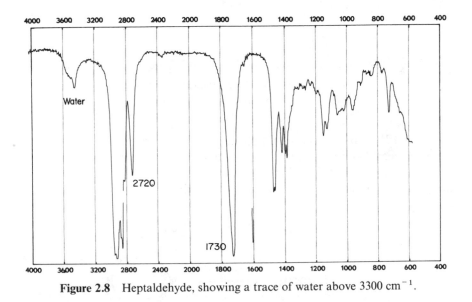

Figure 2.8 Heptaldehyde, showing a trace of water above 3300 cm^{-1}.

diagnostic for aldehydes. Figure 2.8 shows the spectrum of heptaldehyde, slightly wet. It clearly shows the weak aldehyde doublet at 2720 cm^{-1} and a strong carbonyl absorption at 1730 cm^{-1}.

DIKETONES

While α-diketones show no particularly interesting spectral features, β-diketones show two distinct peaks, even if they are symmetrical. Figure 2.9 shows the spectrum of 2,4-pentanedione which exhibits a weak pair of peaks

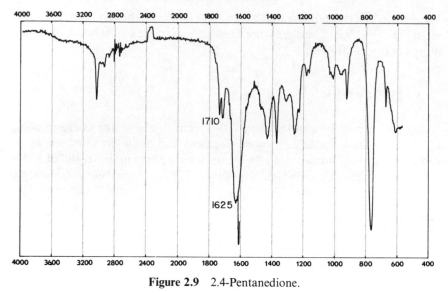

Figure 2.9 2.4-Pentanedione.

at about 1710 cm^{-1} and a much stronger peak at 1625 cm^{-1}. This much
stronger peak lies below our arbitrary unsaturation line and is presumably
due to some sort of conjugation. The most likely conjugation of such a compound is, of course, enolization, and we thus draw the following two likely
structures:

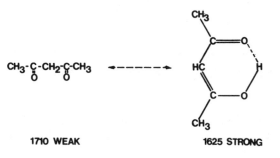

in equilibrium.

CYCLOPROPYL CONJUGATION

Cyclopropanes also show some ability to conjugate with carbonyl compounds, as evidenced by the following series:[8,9]

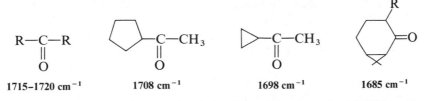

1715–1720 cm^{-1} 1708 cm^{-1} 1698 cm^{-1} 1685 cm^{-1}

It is much as if the cyclopropyl ring itself contained a π-bond. This cyclo-propyl conjugation has been explained by the Walsh model[10] and a similar but more complex bonding theory developed by Coulson and Moffit.[11]

The Walsh model postulates three p-orbitals lying in the plane of the ring and overlapping outside the perimeter of the ring as shown below. Perpen-dicular to the plane of the ring and to these p-orbitals are the three associated sp^2 orbitals on each carbon. Two of these are used to bond with the hydrogens or other substituent groups and the third sp^2 orbital from each carbon is directed into the center of the ring, forming a three-center bond, much as has been observed in various boron compounds. This structure has been verified to some extent by X-ray diffraction studies which indicate that the electron density is substantial outside the nominal perimeter of the ring and nonzero in the center of the ring.

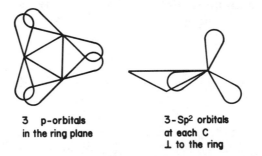

3 p-orbitals
in the ring plane

3-Sp² orbitals
at each C
⊥ to the ring

Because of these p-orbitals in the plane of the ring, the cyclopropyl ring behaves much as if it were a π-system, in conjugation with the carbonyl π-bond, thus leading to increased stability of the carbonyl stretching and a lower energy observed for the stretch.

VIBRATIONAL COUPLING

A number of anhydrides, peroxides, and other diketones show a carbonyl doublet due to vibrational coupling. This is analogous to the *gem*-dimethyl doublets observed because of interaction between the in-phase and out-of-phase stretchings of the methyl groups, and it explains the doublet at 1710

cm^{-1} in 2,4-pentanedione in Figure 2.9. It is also observed in benzoyl peroxide, benzoic anhydride, and malonic esters.

Carboxylic acids, which generally exist as dimers in the solid state, do not show this doublet, presumably because of conjugation:

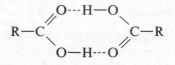

Benzoic acid, for example, shows a carbonyl peak at about 1700 cm^{-1} as a dimer, but in dilute solution it has been observed to show a monomer peak at 1750 cm^{-1}. In carboxylic acids, we see the first example of an —OH group, but as a dimer it is shifted to 2500–2700 cm^{-1}, whereas in the monomer, which is less often observed, we see it around 3500 cm^{-1}. Carboxylic acids also show coupled C—O and O—H stretchings at 1420 and 1300 cm^{-1} and an OH out-of-plane bend at 935 cm^{-1}. Figure 2.10 shows a concentrated CCl_4 solution of benzoic acid, showing the principal peaks.

Dicarboxylic acids also show doublets due to vibrational coupling if zero or one carbon intervenes. Thus we see a carbonyl doublet in oxalic (HOOC—COOH) and maleic acids (HOOC—CH_2—COOH), but not in succinic (HOOC—CH_2—CH_2—COOH).

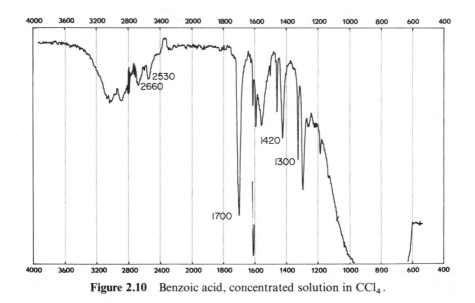

Figure 2.10   Benzoic acid, concentrated solution in CCl_4.

ESTERS

In general, esters come at a higher frequency than ketones since the oxygen destabilizes the carbonyl stretch. For example, the carbonyl peak in acetone is found at about 1720 cm^{-1}, while that of methyl acetate falls at about 1750–1740 cm^{-1}. They usually have an intense band around 1100–1280 cm^{-1} and one or more bands in the overall range of 1000–3000 cm^{-1} formed from the coupling of the C—O stretch to the C—C stretch. Methyl esters often show triplets around 1200 cm^{-1}.

ETHERS

Various alkoxy groups have specific frequencies that can be recognized. These are due to the strong asymmetric C—O—C stretch and are quite characteristic.

Group	ν cm^{-1}
—OCH$_3$	1190
—OCH$_2$CH$_3$	1160, 1100
—O—i-Pr	1175, 1135, 1110
—O—n-Bu	1150, 1125, 1075
—O—t-Bu	770–720, 920–820, 1040–1000, 1200–1155

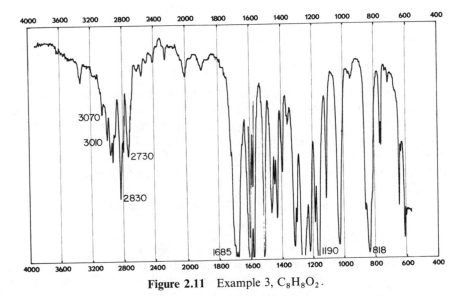

Figure 2.11 Example 3, $C_8H_8O_2$.

In addition, methyl ethers show a strong, characteristic methyl stretching band at 2815–2830 cm^{-1}.

To illustrate analysis of a spectrum showing a number of the features just discussed, let us consider Figure 2.11, a compound having the formula $C_8H_8O_2$. From the formula and from the strong line at 818 cm^{-1}, the compound is clearly aromatic. Furthermore, the pattern between 1600 and 2000 cm^{-1} indicates probable *para* substitution. The strong peak at 1685 cm^{-1} indicates a carbonyl in conjugation with the aromatic ring, and the line at 2730 cm^{-1} is characteristic of aldehydes. Thus one functional group must be an aldehyde group and the compound is a substituted benzaldehyde. The line at 2830 cm^{-1} is characteristic of methoxy groups, as is that at 1190 cm^{-1}, and the compound is therefore *p*-methoxybenzaldehyde or anisaldehyde.

ALCOHOLS

Alcohols are generally easily distinguished by a broad intense peak at 3650–3500 cm^{-1}. Within this range, we can usually predict the substitution of the alcohol as follows:

1°	3639–3633 cm^{-1}
2°	3625–3620 cm^{-1}
3°	3619–3611 cm^{-1}
Phenyl	3611–3603 cm^{-1} much more intense than aliphatic

Alcohols also show additional peaks due to hydrogen bonding into dimers and polymers:

Dimers	3500–3450 cm^{-1}
Polymers	3400–3200 cm^{-1}

All alcohols are partially polymers at room temperature, so most alcohol spectra look like that of ethanol in Figure 2.12, showing a shoulder of free alcohol stretching at 3640 cm^{-1} and a larger hydrogen bonded peak at 3345 cm^{-1}. Note the ethyl rocking frequency at 795 cm^{-1}. The ratio of hydrogen-bonded dimer to monomer can be changed by dilution as shown in Figure 2.12*b*, where the monomer peak is clearly visible in a 5 % solution of ethanol in CCl_4.

Alcohols also exhibit strong C—O stretching bands depending on their substitution

2° and 3° alcohols	1140–1090 cm^{-1}
cyclic alcohols	1065–1015 cm^{-1}
1° alcohols	1060–1025 cm^{-1}

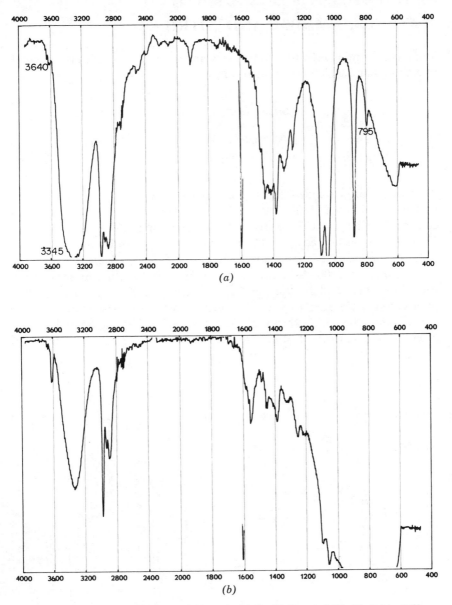

Figure 2.12 Spectra of absolute ethanol: (*a*) liquid film and (*b*) diluted in CCl$_4$.

AMINES, AMIDES, AND NITRILES

The most prominent feature of primary and secondary amines is the N—H stretching band which occurs as follows:

	Stretch (cm^{-1})	Deformation (cm^{-1})
1° amines	3300–3500 doublet	1590–1650 s—m
2° amines	3300–3500 singlet	1650–1510 w

Aromatic amines are identified by two bands in the 1250–1360 cm^{-1} region and one band in the 1180–1280 cm^{-1} region. Tertiary amines are more difficult to recognize, since there are no N—H stretching or deformation bands. They are most easily recognized by treating the compound with mineral acid, isolating the suspected amine salt, and looking for a strong ammonium band in the range of 2200–3000 cm^{-1}. In addition, the methyl amines can be sometimes identified by the N—CH$_3$ stretch at 2760–2820 cm^{-1}. These points are illustrated in Figure 2.13 which shows the IR spectrum of aniline and Figure 2.14 which shows the spectrum of N,N-dimethylaniline. The broad amine salt N—H stretching is shown in Figure 2.15 of anilinium hydrochloride.

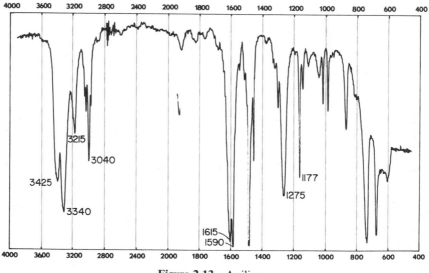

Figure 2.13 Aniline.

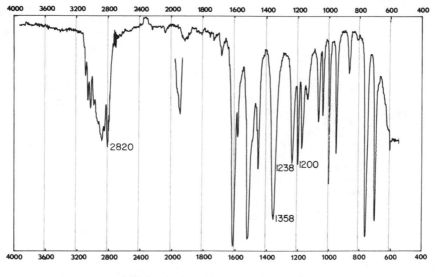

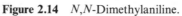

Figure 2.14 *N,N*-Dimethylaniline.

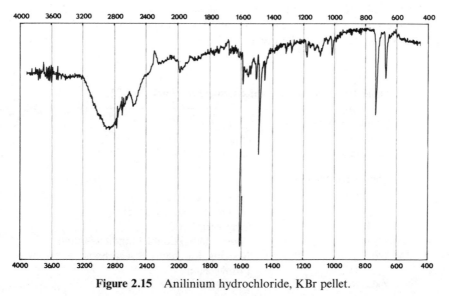

Figure 2.15 Anilinium hydrochloride, KBr pellet.

39

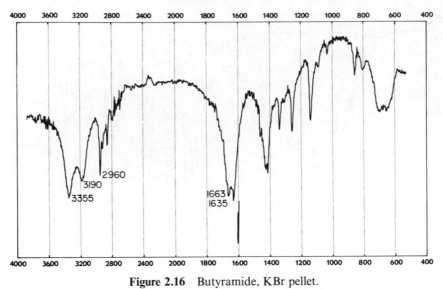

Figure 2.16 Butyramide, KBr pellet.

Amides have spectra much like amines except for the prominent carbonyl peak at $1650-1715$ cm^{-1}, called the amide I band. An additional band, called the amide II band, usually appears just below the carbonyl region. This is illustrated in Figure 2.16 for butyramide.

Nitriles have a unique strong, sharp band in the region of $2200-2300$ cm^{-1} which is characteristic of the C≡N stretch. Conjugated nitriles and aromatic nitriles show peaks near the low end of this range.

Table 2.3 Summary of Major Group Frequencies (cm^{-1})

Mulling Solvent	
2800–2900	Nujol
1500–1300	Nujol
Methyl and Methylene	
2990–2850	—CH$_3$ and —CH$_2$—
1475–1450	Asymmetric bend and scissor
1375	Symmetric bend
1385–1370	*Gem*-dimethyl doublet (and *tert*-butyl)
720	Rocking, for chains > 4 carbons
743–734	Propyl chain
790–770	Ethyl chain

Table 2.3 (*Continued*)

Alkenes

3080	=CH$_2$ Asymmetric stretch
3020	=CHR Asymmetric stretch
1645	RCH=CH$_2$
1655	R$_2$C=CH$_2$
1660	*cis*-RCH=CHR
1675	*trans*-RCH=CHR
1670	R$_2$C=CHR and R$_2$C=CR$_2$

Rings

3103	Cyclopropane asymmetric stretch
3024	Cyclopropane symmetric stretch
3056–2990	Epoxides
1250	Epoxide symmetric ring stretch
950–890	Epoxide asymmetric stretching of ring

Alkynes

3315–3270	≡C—H
2260–2210	C≡C Internal
2140–2100	C≡C External

Nitriles

2300–2200	C≡N Stretch

Aromatic Hydrocarbons

3100–3000	C—H Asymmetric stretch
2000–1650	Benzene substitution pattern region
770–730	Monosubstituted benzenes
710–690	Monosubstituted benzenes
770–735	*ortho*-substituted benzenes
810–750	*meta*-substituted benzenes
710–690	*meta*-substituted benzenes
860–800	*para*-substituted benzenes

Carbonyls

1900–1580	Carbonyl groups
1720–1700	Ketones
1740–1720	Aldehydes
1680–1660	Unsaturated ketones
2900–2700	Aldehyde C—H stretch
1625	β-Diketones

(*Continued*)

Table 2.3 (*Continued*)

Carboxylic Acids

2700–2500	—OH Dimer peak
1420	Coupled C—O stretch in COOH
1300	Coupled C—O stretch in COOH
935	O—H Out-of-plane bend in COOH

Esters

1750–1740	Ester C=O stretch
1280–1000	Ester C—O stretch
1200	Methyl ester triplet

Ethers

1190	—OCH$_3$
2830–2815	Methyl ether band

Alcohols

3639–3633	Primary alcohols (dilute)
3625–3620	Secondary alcohols (dilute)
3619–3611	Tertiary alcohols (dilute)
3611–3603	Aromatic alcohols (dilute)
3500–3450	Alcohol dimers
3400–3200	Alcohol polymers
1140–1090	Secondary and tertiary alcohols
1065–1015	Cyclic alcohols
1060–1025	Primary alcohols

Amines

3500–3300	Primary amines (doublet)
3500–3300	Secondary amines (singlet)
1650–1590	N—H Primary amine deformation
1650–1510	N—H Secondary amine deformation
1360–1250	Aromatic amines
1280–1180	Aromatic amines
3000–2200	Ammonium band
2820–2760	*N*-Methyl amines

Amides

1715–1650	Amide carbonyl (amide I band)
1670–1640	Amide II band

Chlorides

760–540	C—Cl Stretch

Table 2.4   Major Group Frequencies in Descending Order (cm^{-1})

3639–3633	Primary alcohols (dilute)
3625–3620	Secondary alcohols (dilute)
3619–3611	Tertiary alcohols (dilute)
3611–3603	Aromatic alcohols (dilute)
3500–3450	Alcohol dimers
3500–3300	Primary amines (doublet)
3500–3300	Secondary amines (singlet)
3400–3200	Alcohol polymers
3315–3270	≡C—H
3103	Cyclopropane asymmetric stretch
3100–3000	C—H Asymmetric stretch
3080	=CH$_2$ Asymmetric stretch
3056–2990	Epoxides
3024	Cyclopropane symmetric stretch
3020	=CHR Asymmetric stretch
3000–2200	Ammonium band
2990–2850	—CH$_3$ and —CH$_2$—
2900–2800	Nujol
2900–2700	Aldehyde C—H stretch
2830–2815	Methyl ether band
2820–2760	N-Methyl amines
2700–2500	—OH Dimer peak
2300–2200	C≡N Stretch
2260–2210	C≡C Internal
2140–2100	C≡C Internal
2000–1650	Benzene substitution pattern region
1900–1580	Carbonyl groups
1750–1740	Ester C=O stretch
1740–1720	Aldehydes
1720–1700	Ketones
1715–1650	Amide carbonyl (amide I band)
1680–1660	Unsaturated ketones
1675	trans-RCH=CHR
1670	R$_2$C=CHR and R$_2$C=CR$_2$
1670–1640	Amide II band
1660	cis-RCH=CHR
1655	R$_2$C=CH$_2$
1650–1590	N—H Primary amine deformation
1650–1510	N—H Secondary amine deformation
1645	RCH=CH$_2$
1625	β-Diketones

(*Continued*)

Table 2.4 (*Continued*)

1500–1300	Nujol
1475–1450	Asymmetric bend and scissor
1420	Coupled C—O stretch in COOH
1385–1370	*Gem*-dimethyl doublet (and *tert*-butyl)
1375	Symmetric bend
1360–1250	Aromatic amines
1300	Coupled C—O stretch in COOH
1280–100	Ester C—O stretch
1280–1180	Aromatic amines
1250	Epoxide symmetric ring stretch
1200	Methyl ester triplet
1190	—OCH$_3$
1140–1090	Secondary and tertiary alcohols
1065–1015	Cyclic alcohols
1060–1025	Primary alcohols
950–810	Epoxide asymmetric stretching of ring
935	O—H Out-of-plane bend in COOH
860–800	*para*-substituted benzenes
810–750	*meta*-substituted benzenes
790–770	Ethyl chain
770–730	Monosubstituted benzenes
770–735	*ortho*-substituted benzenes
760–510	C—Cl stretch
743–734	Propyl chain
720	Rocking, for chains > 4 carbons
710–690	Monosubstituted benzenes
710–690	*meta*-substituted benzenes

PROBLEMS IN COMPOUND IDENTIFICATION

The following spectra of unknown compounds are to be identified from the IR spectrum and the supplied empirical formula. We shall see in Chapter 9 that this formula can be easily obtained from the mass spectrum. To proceed efficiently, you should first identify the number of degrees of unsaturation of the number of "rings plus double bonds." While this may often be apparent by inspection, the following analytical formula may be helpful in confirming your first impression. The number of rings plus double bonds R is given by

$$R = (n + 1) - \frac{(m - t)}{2}$$

where n = the number of tetravalent atoms (usually the number of carbons)

m = the number of monovalent atoms (usually H or halogens)

t = the number of trivalent atoms (usually N)

Note that divalent atoms such as O and S are ignored. For example, the formula C_7H_7NO has $n = 7$, $m = 7$, $t = 1$, and

$$R = (7 + 1) - \frac{(7 - 1)}{2} = 5.$$

The compound thus has five rings plus double bonds. The formula was taken from benzamide, which has one ring and four double bonds.

In the problem spectra, relevant peaks are indicated by frequency. To make best use of these problems, study them carefully and attempt to eliminate all unlikely structures before consulting the answers.

Problem 2.1 C_6H_{14}

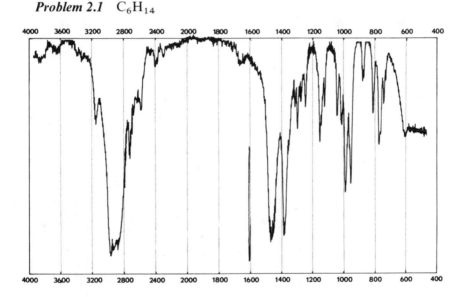

Problem 2.2 C_8H_{18}

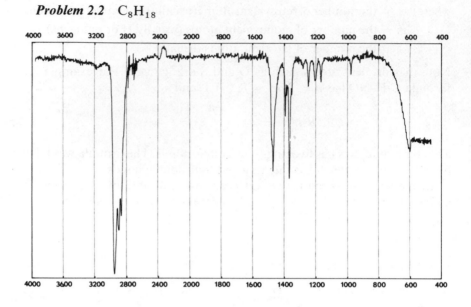

Problem 2.3 C_7H_5N

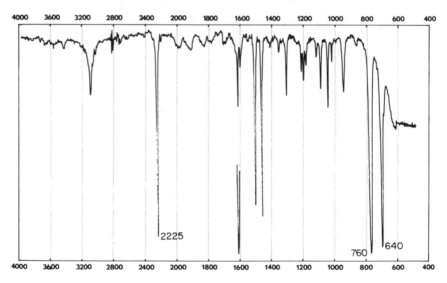

Problem 2.4 $C_4H_{10}O$

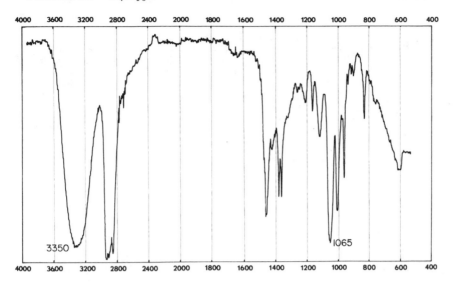

Problem 2.5 $C_6H_{15}N$

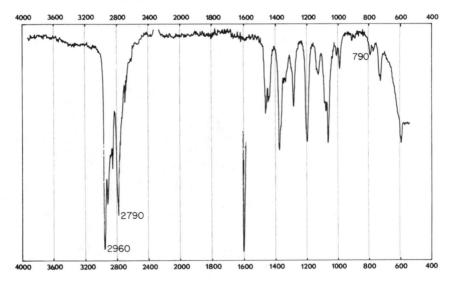

Problem 2.6 $C_8H_8O_2$

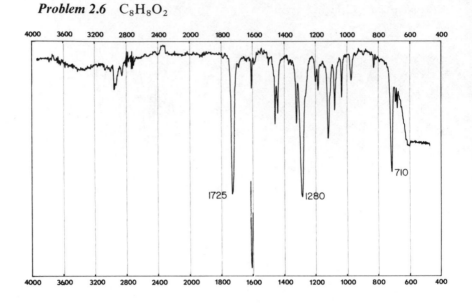

Problem 2.7 $C_9H_{10}O$

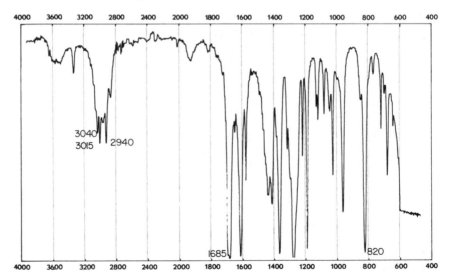

Problem 2.8 C_6H_6ClN

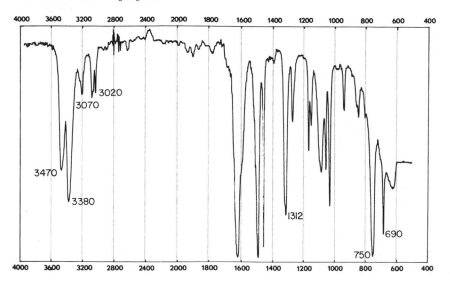

Problem 2.9 $C_7H_8O_2$

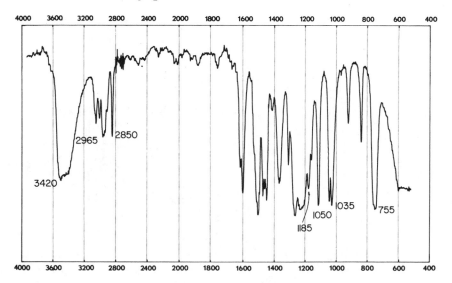

Problem 2.10 C_7H_5OCl

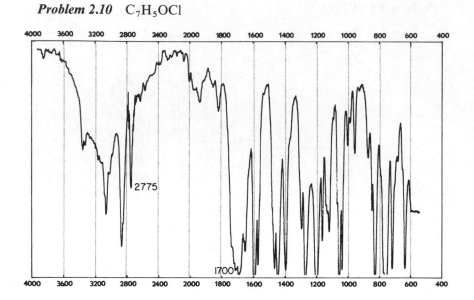

Problem 2.11 $C_6H_{10}O$

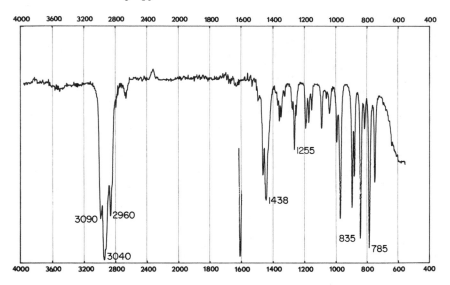

Problem 2.12

The IR spectrum of an unknown compound contains a doublet in the $C{=}O$ stretching region. Suggest at least four reasons why this might be true.

Problem 2.13

Oki and Iwamura (*Tetrahedron Lett.*, **1973** (41), 4003–4006) reported that they observed two bands in a number of simple alcohols in a 0.001 *M* solution in CCl_4 at 3633 and 3619 cm^{-1}. They attributed these bands to different rotamer populations. Account for the differences in intensity.

	3633.7 cm^{-1} Intensity	3619 cm^{-1} Intensity
Ethanol	1.0	0.56
Propanol	1.0	0.38
Isobutanol	1.0	0.50
Cyclopropylmethanol	1.0	1.1

Problem 2.14

How could you distinguish the IR spectra of ethyl benzoate and phenyl acetate?

BIBLIOGRAPHY

Introductory Texts

Conley, R. T., *Infrared Spectroscopy*, 2nd ed., Allyn and Bacon, Boston, 1972.

Dyer, John R., *Applications of Absorption Spectroscopy of Organic Compounds*, Prentice-Hall, Englewood Cliffs, N.J., 1965.

Lambert, J. B., H. F. Shurvell, L. Verbit, R. G. Cooks, and G. H. Stout, *Organic Structural Analysis*, Macmillan, New York, 1976.

Nakanishi, K. and P. H. Solomon, *Infrared Absorption Spectroscopy*, Holden-Day, San Franciso, 1978.

Scheinmann, F., Ed., *An Introduction of Spectroscopic Methods for the Identification of Organic Compounds*, Pergamon Press, New York, 1970.

Advanced Texts

Alpert, N. L., W. E. Keiser, and H. A. Szymanski, *IR—Theory and Practice of Infrared Spectroscopy*, Plenum Press, New York, 1973.

Bellamy, L. J., *The Infrared Spectra of Complex Molecules*, 2nd ed., Methuen, London, 1958.

Rao, C. N. R., *Chemical Applications of Infrared Spectroscopy*, Academic Press, New York, 1963.

Sources of Additional Problems

Dyer, John R., *Organic Spectral Problems*, Prentice-Hall, Englewood Cliffs, N.J., 1972.

Mason, A. N., A. Bhati, and J. Cast, *Spectroscopic Exercises in Organic Chemistry*, Halsted-Wiley, New York, 1973.

Shapiro, R. H., and C. H. Depuy, *Exercises in Organic Spectroscopy*, 2nd ed., Holt, Rinehart and Winston, New York, 1977.

Collections of Spectra

Colthup, N. B., L. H. Daly, and S. E. Wiberly, *Introduction to Infrared and Raman Spectroscopy*, Academic Press, New York, 1964.

Pouchert, C. J., "The Aldrich Library of Infrared Spectra," Aldrich Chemical Co., 940 W. St. Paul Ave., Milwaukee, Wisc. 53233.

"Sadtler Standard Infrared Spectra," Sadtler Research Laboratories, Inc. 3316 Spring Garden St., Philadelphia, Pa. 19104.

Szymanski, H. A., *Interpreted Infrared Spectra*, 3 vols., Plenum Press, New York, 1964.

REFERENCES

1. E. K. Plyler and N. Acquista, *J. Res. Natl. Bur. Stand.*, **43**, 37 (1949).
2. S. E. Wiberly, S. C. Bunce, and W. H. Bauer, *Anal. Chem.*, **32**, 217 (1960).
3. C. W. Young, R. B. Duvall, and N. Wright, *Anal. Chem.*, **23**, 709 (1951).
4. R. N. Hazeltine and K. Leedham, *J. Chem. Soc.*, **1952**, 3483.
5. A. H. Nielson et al., *J. Chem. Phys.*, **20**, 596 (1952).
6. R. N. Jones, W. F. Forbes, and W. A. Muellen, *Can. J. Chem.*, **35**, 504 (1957).
7. C. N. R. Rao, G. K. Goldman, and J. Ramachandran, *J. Indian Inst. Sci.*, **43**, 10 (1961).
8. S. E. Wiberly and S. C. Bunce, *Anal. Chem.*, **24**, 623 (1952).
9. R. J. Mohrbacher and N. H. Cromwell, *J. Am. Chem. Soc.*, **79**, 401 (1957).
10. A. D. Walsh, *Trans. Faraday Soc.*, **45**, 179 (1949).

CHAPTER THREE
INTRODUCTION TO
NUCLEAR MAGNETIC RESONANCE

The first indication of the phenomenon of nuclear magnetic interactions came in 1924 when Pauli[1] suggested that the hyperfine structure of atomic spectra might be due to interaction of the magnetic moments of individual nuclei with the moments of electrons. Magnetic resonance was not directly observed, however, until 1946 when Purcell[2] (Harvard) found hydrogen resonance in paraffin wax and Bloch[3] (Stanford) found hydrogen resonance in water. For these efforts they shared the Nobel prize.

Nmr became of interest to chemists only after Knight[4] reported in 1949 that the precise frequency of energy absorption depends on the chemical environment of hydrogens. Since then nmr has progressed so that it is not only the first or second measurement taken on a compound, but is also commonly used in both graduate research and undergraduate instruction. It has the advantage of being a totally nondestructive technique, and with modern routine instruments spectra can be obtained as fast as or faster than high resolution IR spectra. Furthermore, solids are much easier to study than in IR since a wide variety of solvents have no absorptions in the ranges that are of interest.

The simplest possible introductory description of the nmr phenomenon points out that a nucleus can be considered to be a spinning charged particle which thus has an associated magnetic moment. In the absence of any magnetic field, this moment can be pointing in any direction, but when the sample is placed in a magnetic field, the magnetic moments can align themselves either with or against the field. The higher energy state, against the field, is somewhat less populated than the lower energy state, with the field, and the nuclei can be promoted from the lower to the higher state by the application of radiofrequency energy. It is the absorption of this energy which we observe in the nmr experiment.

Many common nuclei have no magnetic moment and thus will not exhibit any magnetic resonance absorptions. These include ^{12}C, ^{16}O, and ^{32}S since they have a nuclear spin of 0. In fact, any isotope with an even mass number and even atomic number has a spin of zero. Some of the important nuclei with a nuclear spin of $\frac{1}{2}$ are:

$$^1\text{H}, \, ^3\text{H}, \, ^{13}\text{C}, \, ^{15}\text{N}, \, ^{19}\text{F}, \quad \text{and} \quad ^{31}\text{P}.$$

Other nuclei that are of interest to us have somewhat larger nuclear spins:

$I = 1$: ^2H, ^{14}N.

$I > 1$:	^{10}B,	^{11}B,	^{17}O,	^{23}Na,	^{27}Al,	^{35}Cl,	^{59}Co.
I:	3	$\frac{3}{2}$	$\frac{5}{2}$	$\frac{3}{2}$	$\frac{5}{2}$	$\frac{3}{2}$	$\frac{7}{2}$

CLASSICAL DESCRIPTION OF NMR

If we regard nuclei as spinning charged particles, their nuclear spin has an angular momentum **p** that is quantized in units of $h/2\pi$:

$$\mathbf{p} = \frac{Ih}{2\pi} = I\hbar. \tag{3.1}$$

Here I is an integer or half integer, called the nuclear spin quantum number, or just the *nuclear spin*. All such nuclei have a magnetic moment μ which is proportional to the nuclear spin. The proportionality constant γ is called the *magnetogyric ratio*.

$$\boldsymbol{\mu} = \gamma \mathbf{p}. \tag{3.2}$$

If we place a sample containing such nuclei in a magnetic field, the magnetic interaction between the applied magnetic field H_0 and μ generates a torque

Figure 3.1 Precession of a nuclear magnet μ about the applied magnetic field $\mathbf{H}_0$.

that tends to tip the magnetic moment toward the applied field. At the same time, since the nucleus is spinning, it tends to *precess* or wobble around H_0 as shown in Figure 3.1, much as a toy gyroscope precesses around the Earth's gravitational field. Now the rate of precession around the magnetic field is proportional to the applied field and the magnetogyric ratio

$$v = \frac{\gamma H_0}{2\pi} \tag{3.3}$$

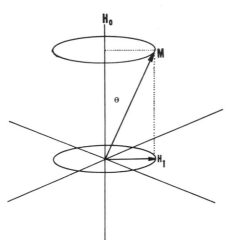

Figure 3.2 Precession of a group of nuclear magnets **M** about the applied field H_0 in phase with the magnetic component of an applied rf field H_1.

and this precession frequency v is called the *Larmor* frequency. Since each element has its own magnetogyric ratio, we find that each nucleus precesses at a different characteristic Larmor frequency.

While the frequency of precession is fixed, the angle of precession is related to the applied energy. If we apply a small rotating field in the x—y plane, we can vary the precession angle, θ, and tip the magnetic vector μ, so that it has a greater component in the x—y plane. While a rotating magnetic field sounds like an extremely difficult physical problem, a radiofrequency field applied at the right angle will have a rotating magnetic component in the x—y plane that will induce the magnetic moment μ to tip to a different angle. This is illustrated in Figure 3.2. It is necessary that this rotation frequency be exactly the same as the Larmor frequency for the nuclei we wish to excite or there will be no absorption of energy or *resonance*, and thus we usually vary either the frequency or the magnetic field until resonance is observed. A plot of some common resonance frequencies for various nuclei is shown in Figure 3.3 below for a 14.1 kG (1.41 tesla) field.

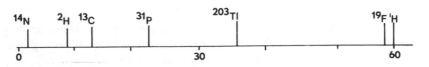

Figure 3.3 Absorption frequencies of some common nuclei at a 14.1 kG field (1.41 tesla).

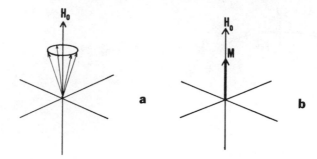

Figure 3.4 (*a*) Precession of a group of nuclear magnets about the applied field H_0. (*b*) Resultant magnetization of the group in (*a*) pointing along the applied field H_0.

OBSERVATION OF RESONANCE

Thus far we have shown how application of a radiofrequency field that has a rotating magnetic component will tip the magnetic moment of a particular nucleus from its precession angle to a new angle closer to the x—y plane. When we do this the magnetic moment has a greater component in the x—y plane perpendicular to the applied field H_0, and this component can be observed by a receiver coil placed in this plane. The component will also be a rotating magnetic field, and it will induce a current in the receiver coil that can then be amplified and plotted on a chart recorder, indicating that resonance has occurred.

It might seem at first that since the magnetic moment μ is precessing even before the radiofrequency field is applied, we should observe some magnetic component in the x—y plane at all times. This does not occur, however, because the various μ's of the many molecules making up the sample are not all rotating in phase, and their magnetic moments cancel out, leaving a net resultant magnetic moment **M** parallel with the field. This is illustrated in Figure 3.4. It is only when a rotating magnetic field is applied that the various magnetic moments all begin rotating together and become phase coherent, pointing in the same direction at the same time, as shown in Figure 3.2.

RELAXATION OF THE EXCITED NUCLEI

Once the nuclei have absorbed this energy and the varying frequency has moved on, the nuclei clearly do not remain at the new nonequilibrium angle, which is a higher energy state than the usual precession angle. Instead, they relax by two processes to reach energy equilibrium: *spin–lattice* relaxation and *spin–spin* relaxation.

Spin–lattice (or longitudinal) relaxation is a transition to a lower energy state by the various spins interacting with the environment physically, that is, with the *lattice*. The decay is exponential in nature:[5]

$$n - n_{eq} = (n - n_{eq})_{t=0} \exp \frac{-t}{T_1} \tag{3.4}$$

where T_1 is called the spin–lattice relaxation time, and is usually between 10^{-1} and 10^2 sec for liquids and solutions, n is the population in the excited state, n_{eq} is the equilibrium population, and the quantity $(n - n_{eq})_{t=0}$ is the population difference before relaxation occurs. Spin–lattice relaxation can be observed by sweeping through a resonance line in a spectrum at high power and then sweeping back through it again; the second sweep will produce a much less intense signal since little relaxation has yet occurred. This phenomenon is known as *saturation*. When an nmr line is saturated, the populations of the upper and lower states are equal and upward and downward transitions between the energy levels are equally probable. The spin–lattice relaxation time T_1 thus defines with how much power and how often you can sweep through a spectrum.

Spin–spin (or transverse) relaxation is the interaction of the nuclei in question with nearby magnetic moments of other species, causing some local variation in the field that a particular nucleus " sees." Thus all the nuclei that are theoretically chemically equivalent do not resonate at exactly the same frequency, and nmr lines are not infinitely sharp, but have some line width associated with them. Because of these local fields, some nuclei will be precessing at a frequency slightly different from those at the center of the nmr resonance line. They therefore will not stay in phase with the rest of the magnetic moments, precessing either slower or faster, and will contribute to the reduction of phase coherence among the packet of magnetic moments. When the moments are no longer all pointing in the same direction, neither will their components in the $x-y$ plane, and the vectors will tend to cancel each other, leading to less net absorption. The spin–spin relaxation time T_2 is thus related to the line width observed at half-height $v_{1/2}$:

$$T_2 = \frac{1}{\pi v_{1/2}}.$$

Note that the narrower the line width the *longer* T_2 will be.

DISTRIBUTION OF SPINS

If we consider a population of spin-$\frac{1}{2}$ nuclei having the two energy levels $-\frac{1}{2}$ and $+\frac{1}{2}$, we can designate the populations in the two levels by the Boltzmann distribution:

$$\frac{n_+}{n_-} = \exp\frac{(-\Delta E)}{kT} \tag{3.5}$$

where n_+ is the number in the upper level, n_- is the number in the lower level, k is the Boltzmann constant, and T is the Kelvin temperature. The energy is proportional to the magnetogyric ratio and the applied field:

$$E = \frac{-\gamma h H_0}{2\pi} = -\gamma \hbar H_0 \tag{3.6}$$

and we can then write

$$\frac{n_+}{n_-} = \exp\frac{(-\gamma \hbar H_0)}{kT}. \tag{3.7}$$

From this and the fact that $e^{-x} = 1 - x$ for $x \ll 1$, Becker[5] has shown that

$$\frac{n_- - n_+}{n_-} = \frac{2\mu H_0}{kT} \tag{3.8}$$

and thus for ^{1}H at 14 kG, the excess population in the lower level is only 1×10^{-5}. The energy transition that is observed is thus a very small one and since only a small net excess of nuclei is actually promoted, nmr has the reputation of being a relatively insensitive technique. Not only ^{1}H, but ^{13}C, which has a natural abundance of 1.1%, is routinely observed, which is a tribute to the designers of modern spectrometers. In actual fact, ^{1}H nmr on the microgram scale[6] is perfectly possible with today's instruments.

INSTRUMENTATION OF SWEPT NMR SPECTROMETERS

The simplest nmr spectrometer consists of a large permanent or electromagnet, a transmitter, a receiver, and a facility for sweeping (varying) either the frequency of the transmitter or the magnetic field and for sweeping across a flat-bed recorder simultaneously. The magnet homogeneity is achieved by restricting the sample volume to a small area near the center of the magnet poles, using polished, optically flat pole caps and carefully thermostatting the magnet. Additional small electromagnet coils are embedded in the magnet pole pieces and even in the probe itself to allow slight variations in the local field to achieve maximum field homogeneity or uniformity. These coils are called *shim* coils and add on slight magnetic field gradients under operator control. Finally, the nmr sample tubes are carefully produced to be axially symmetric and are spun in the spectrometer to further average out local inhomogeneities.

There are several levels of locking provided to insure that the magnetic field remains stable and reproducible. The first is a flux stabilizer that simply detects gross changes in the magnetic field and then sweeps the magnet power supply to restore the correct field.

A more sensitive system for field stabilization involves the detection of the resonant frequency of a known species either sealed in the probe (external lock) or actually part of the material in the sample tube (internal lock). This signal is monitored by the locking circuit and if the resonance starts to drift the field is corrected to bring the signal back into resonance. This is accomplished by observing the signal not as a peak but as a first derivative (dispersion mode) line, so that a drift in one direction will generate a positive voltage, a drift in the other will generate a negative voltage, and a centered peak will generate zero voltage.

SAMPLE PREPARATION

The typical ^{1}H or proton nmr spectrum (pmr) is obtained from a sample in a 5 mm thin-walled glass tube containing about 0.4 ml of sample. If sample quantity is severely restricted, capillary tubes are available with a small spherical cavity at the bottom, thus concentrating all the available sample in a small area within the receiver coils. Sample concentrations for routine work can be as low as 0.01 M but concentrations greater than 0.2 M are preferred for good signal-to-noise ratio.

Liquid samples are seldom run as neat liquids, since their greater viscosity will lead to broader lines from decreased T_2s. Instead, the liquids and solids are dissolved in a suitable solvent that does not show any peaks in the region of interest. With the advent of lower cost deuterated solvents, this usually means only trace amounts of any proton resonances. Common solvents include CCl_4, $CDCl_3$, D_2O, acetone-d_6, and DMSO-d_6.

It is important that no undissolved solids or impurities be present in the final sample solution, since such particles cause local field variations, which can lead to extreme broadening of lines in the nmr spectrum. The sample solution should therefore never be prepared in the actual nmr tube, but in a small test tube so that heating, agitation, and filtration are possible before transferring the sample to the nmr tube.

The nmr tubes are filled about 2–3 cm deep with the sample solution and capped with a pressure cap. The caps commonly supplied with nmr tubes have a fairly wide shoulder, which causes more wobble than necessary when the tubes are spun; we therefore recommend that the caps be replaced with the shoulderless polyethylene pressure caps available from Wilmad.

CLEANING OF NMR TUBES

It is necessary that the nmr tubes be scrupulously clean and dry before use to prevent contaminants from introducing spurious spectral lines. Furthermore, small traces of paramagnetic impurities will cause severe broadening of resonance lines. While new nmr tubes are usually quite clean, the method of cleaning old tubes before reuse has become something of a problem in many research groups. If the previous sample was a dilute solution of some highly soluble substance, there is no problem in rinsing out the tube with a stream of acetone or chloroform from a wash bottle. An nmr sample tube washer, which will do a good job of getting solvent to the bottom of the tube, is available from Norell Chemical Company (Landisville, N.J.). Alternate methods include attaching a 12-in., 18-gauge syringe needle via tubing to a wash bottle or faucet to rinse out the tube.

When the previous contents of the tube are unknown, however, or when they have decomposed and left solid materials behind, the cleaning of the tubes becomes more difficult. If the materials are somewhat soluble, Barcza has suggested an apparatus consisting of a wide tube connected to a pot of solvent at the bottom and a condenser on the top and containing a basket of vertical nmr tubes, open end down supported by a wire screen. He reports that refluxing solvent in this apparatus for a day or so will clean most tubes.

If it is necessary to use various cleaning solutions, aqua regia (three parts HCl to one part HNO_3) is the solution of choice. Chromic acid should be avoided since the paramagnetic impurities it will introduce by precipitation will be difficult to remove. A second soaking of the tubes in a saturated solution of trisodium phosphate followed by rinsing with water and acetone will usually remove all materials from the tubes.

Finally, ultrasonic cleaning devices, which are reported to clean tubes well but which are not inexpensive, are available from various suppliers.

PRESENTATION OF NMR SPECTRA

Nmr spectra are usually plotted from low field to high field going from left to right. If the frequency is swept, it will be from high frequency to low frequency from left to right, but it is still referred to as having a low-field and high-field end.

If the peak separations are to be measured in an nmr spectrum, it is necessary to have a zero point to measure *from*. This reference has been agreed upon as the peak produced by tetramethylsilane or TMS. This sharp singlet has the advantage of being at the right end of most spectra; only a very few compounds show peaks further upfield than TMS. The compound TMS has the further advantage of being quite volatile and easy to remove from the solution when the sample is to be recovered. It is soluble in most nmr solvents other than D_2O and DMSO and since it has such a large number of hydrogens per unit weight, only a small amount of TMS is required.

Since TMS is so volatile, it is seldom actually added to the sample solution, but is instead added to a bottle of solvent to make a 1 % solution of TMS. This will have a much lower vapor pressure than pure TMS and can be kept at room temperature in tightly closed bottles. By contrast, TMS itself, which boils at 26.5°C, should always be kept refrigerated and added to solutions with a refrigerated dropper. It should never be added directly to an nmr tube when the solution contains a dissolved solid, however, because it nearly always causes the formation of a precipitate, which can cause severe line broadening.

For samples to be run in D_2O or DMSO either 3-(trimethylsilyl)-propane-sulfonic acid, sodium salt (DSS), or 3-(trimethylsilyl)-tetradeuterosodium propionate (TSP) are used. Other compounds that have sharp singlets and can be used as secondary references include cyclohexane, benzene, acetone, *tert*-butanol, dioxane, and acetonitrile. Generally, the reference compound is mixed with the sample solution. When this is not possible, the reference is placed in a small sealed capillary and held rigidly in place inside the sample tube.

MEASUREMENT OF CHEMICAL SHIFTS

If we made up a solution of *tert*-butanol, cyclohexane, acetone, dioxane, water, and benzene, added some TMS as a reference, and obtained its nmr spectrum in a 14.1 kG field at around 60 MHz, we would find that we would observe peaks 83.4, 97.8, 112.3, 228, 308.4, and 429 Hz from TMS. This is illustrated in Figure 3.5a. If we took this same sample and obtained its nmr spectrum at 23.48 kG at around 100 MHz, we would find that the peaks would be observed at 139, 163, 187, 380, 514, and 715 MHz from TMS. Thus the chemical shift in Hz seems to be dependent on the applied field. Recalling Eq. 3.3,

$$v_0 = \frac{\gamma}{2\pi} H_0, \qquad (3.3)$$

we find that this is exactly what we have predicted. Since a large number of

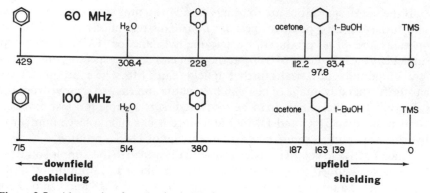

Figure 3.5 Absorption frequencies in Hz from TMS for some common compounds in (*a*) 60 MHz and (*b*) 100 MHz nmr spectrometers.

magnetic fields are available in various commercial spectrometers, as shown in Table 3.1, we would like to have some way of normalizing these frequency separations or *chemical shifts* to a set of numbers that is independent of the applied field and frequency.

Since the chemical shift separation is linear with frequency (or field) we need only divide through by one or the other. The convention used is

$$\text{chemical shift (ppm)} = \frac{\text{shift (Hz) from reference}}{\text{spectrometer frequency (MHz)}}$$

Table 3.1   Proton Resonance Frequencies at Various Magnetic Fields

Frequency (MHz)	Field (kG)
60	14.092
90	21.138
100	23.486
180	42.276
200	46.973
220	51.671
250	58.716
270	63.414
360	84.552
600	140.920

Since the units are Hz/MHz, the chemical shift is referred to as *parts per million* or *ppm*. Thus the chemical shift of *tert*-butanol becomes

$$v = \frac{83.4 \text{ Hz}}{60 \text{ MHz}} = 1.39 \text{ ppm.}$$

The symbol (δ) has been adopted to represent chemical shift, and the current convention is that it precedes the number:

$$v = \delta 1.39.$$

This δ scale, then, increases in the positive direction as we go downfield (left) or increase in frequency, and any peaks further upfield than TMS will have a *negative* chemical shift on this scale.

An additional scale, called the τ *scale*, is defined as follows:

$$\tau = 10 - \delta.$$

This makes TMS have a chemical shift of $\tau 10$ and makes all peaks below it have positive, but decreasing, shifts. Thus benzene would have a chemical shift of

$$v = \frac{429 \text{ Hz}}{60 \text{ MHz}} = \delta 7.15 = \tau 2.85.$$

On the τ scale, peaks more than 10 ppm below TMS have negative shifts and those above TMS have shifts greater than $\tau 10.0$. This scale is now seldom used but occurs commonly in papers of the 1960s. Additional confusion occurs because there have been a few chemists who refuse to conform to either of these conventions and reverse the signs of the scale to suit themselves. In these cases, the correct values may only be obtained by careful reading of the article.

Note that the extreme accuracy of nmr measurements allows us to examine peaks only a fraction of a Hz apart in the MHz range. This sort of resolution has been compared to the optical ability to resolve two black cats seated 6 ft apart on the moon! No mention was made, however, of the diameter of the space suits the cats were wearing.

REFERENCES

1. W. Pauli, Jr., *Naturwiss.*, **12**, 741 (1924).
2. E. M. Purcell, H. C. Torrey, and R. V. Pound, *Phys. Rev.*, **69**, 37 (1946).
3. F. Bloch, W. W. Hansen, and M. Packard, *Phys. Rev.*, **69**, 127 (1946).
4. W. D. Knight, *Phys. Rev.*, **76**, 1259 (1949).
5. E. D. Becker, "High Resolution Nmr," Academic Press, New York, 1969, pp. 20–23.
6. T. C. Farrar, in *Transform Techniques in Chemistry*, P. Griffiths, Ed., Plenum Press, New York, 1978.

CHAPTER FOUR
CHEMICAL INTERPRETATION OF
PROTON NUCLEAR MAGNETIC
RESONANCE SPECTRA

CHEMICAL SHIFTS

We have seen that the exact peak positions in a proton nmr spectrum vary considerably with the type of compound involved. In fact, if we look at the schematic spectrum in Figure 4.1, we see that we can make some generalizations about the chemical shifts observed and the functional groups they represent.

We find that chemical shifts are closely related to electron density around the carbon bearing the hydrogens and that the more electron density there is, the further upfield the resonance occurs. We say, then, that those hydrogens having greater electron density are more *shielded* and that those having less electron density near them are more *deshielded*. For example, in Table 4.1 we examine a series of chemical shifts of the methylene hydrogens in the compound CH_3CH_2X and find that they correlate well with the *electronegativity* of X.[1] We also find that the more acidic protons are found at lower fields. For example, phenol —OH hydrogens are found 4 ppm below those of ethanol.

There are some chemical shifts that are not predictable solely from electronegativity effects, such as those of the alkenes and alkynes:

$$CH_3-CH_3 \qquad CH_2{=}CH_2 \qquad H-C{\equiv}C-H$$
$$\delta 0.9 \qquad\qquad \delta 5.0 \qquad\qquad \delta 2.3$$

We would have expected that there would be the least electron density around the alkyne hydrogens, but we actually find that they are upfield from the alkene hydrogens. We attribute this anomaly to the *anisotropy* of the C≡C

64

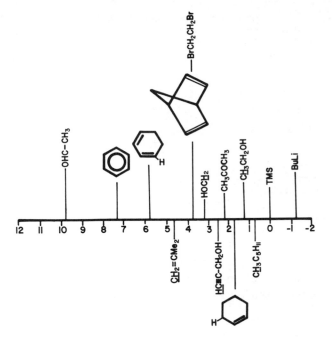

Figure 4.1 Chemical shifts in ppm for a number of common organic compounds.

bonds, or in other words, the nonuniform distribution of magnetic fields in the molecule, which affects the chemical shift. In the case of acetylene, the two perpendicular pairs of π-bonds form a cylinder about which electrons can flow, thus causing some additional shielding of the acetylene hydrogens, as shown in Figure 4.2. The electrons circulate around the π-bonds so as to oppose the applied field and increase the shielding of the hydrogens.

Table 4.1 Chemical Shifts of CH_3CH_2X

X	$\delta\ CH_2$	χ
$SiEt_3$	0.6	1.9
H	0.75	2.2
CEt_3	1.3	2.5
NEt_2	2.4	3.0
OEt	3.3	3.5
F	4.0	4.0

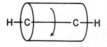

Figure 4.2 The shielding cylinder of acetylene. The acetylene protons are in this cylinder and are thus more shielded than similar vinyl protons.

In carbonyl groups, we also see extensive deshielding of aldehyde protons because of a similar anisotropic effect (Figure 4.3):

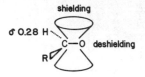

Figure 4.3 Shielding and deshielding regions for aldehydes, showing the deshielding of the aldehyde proton.

Finally, in aromatic rings, we see that a " ring current " is established around the ring by the circulation of electrons, and that the aromatic hydrogens do not lie within this shielding area, but are deshielded (Figure 4.4).

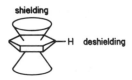

Figure 4.4 Shielding and deshielding regions for benzenes, showing the deshielding of the aromatic protons.

These anisotropic shieldings have been well documented in compounds constructed to show the aromatic shielding cone, such as [18]-annulene[2] (**1**) and protoporphyrin-9[3] (**2**).

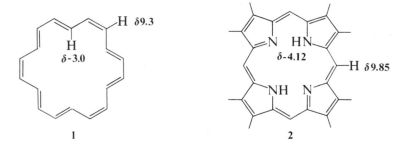

1   2

Another example, less dramatic but also important, is in [10]-para-cyclophane (**3**), where the protons directly above the ring (ε-protons) resonate at $\delta0.3$. Furthermore, cyclopropyl hydrogens are also found at very high field, usually around $\delta0.2$ because of the cyclopropane-ring anisotropy.

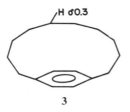

3

Bearing in mind these electronic and anisotropic cases, however, we can make a few generalizations about chemical shifts that are useful in interpreting spectra. They are shown in Table 4.2 below.

Let us now look at a few simple spectra. Figure 4.5 shows the nmr spectra of toluene and *p*-xylene. The chemical shifts of the two spectra are nearly

Table 4.2   Chemical Shifts of Common Proton Groups

Functional Group		δ
Organometallic	RCH_2M	-1.2–0.1
Cyclopropyl	$\triangleright$—H	0.2
Methyl	—CH_3	0.9
Methylene	—CH_2—	1.3
Tertiary	R_3C—H	1.5
Allylic	—C=C—CH_2—	1.7
Alkyne	C≡C—H	2–3
Benzylic	Ar—CH_2—	2.3–3
Vinyl	C=C—H	4.5–6.0
Aromatic	Ar—H	6–8.5
Amino	$\{$ $RC\underline{H}_2NH_2$	2.0–2.8
	R—NH_2	1–5
Alkyl halides	R—CH_2—X	2–4
Ketones	R—CH_2—CO—R	2–2.7
Alcohols	$\{$ $RC\underline{H}_2$—OH	3.4–4
	R—OH	1–5
Ethers	R—O—CH_2—R	3.3–4
Esters	R—CO—OCH_2—R	3.7–4.1
	RCH_2—COOR	2–2.2
Aldehydic	RCO—H	9–10
Carboxylic acids	R—COOH	10–12
Enolic	C=C—OH	15–17

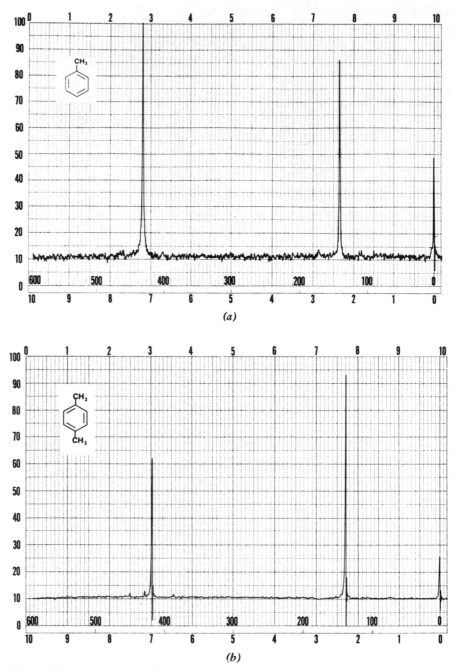

Figure 4.5 Nmr spectra of (*a*) toluene and (*b*) *p*-xylene, showing the difference in peak areas with relative numbers of equivalent protons (5 : 3 versus 4 : 6). This and all similar spectra courtesy of Aldrich Chemical Co., Milwaukee, Wisconsin.

68

identical for the methyl protons and those on the aromatic ring. We should note here that thus far we have indicated that chemically identical protons give identical chemical shifts. The actual requirement is that the nuclei be *magnetically* equivalent, which is more restrictive in some cases, as we will see on p. 74. In the cases of monoalkyl-substituted benzenes and some other benzenes the remaining aromatic protons are so similar in magnetic shielding that they give a single broad line.

In these two spectra, however, there is a substantial difference: the areas under the aromatic peak and the alkyl peaks are different. In the spectrum of toluene (Figure 4.5*a*), the areas are about 5:3 and in the spectrum of *p*-xylene, the areas are about 2:3 or 4:6. In other words, the areas under the peaks are proportional to the number of protons of that kind in the compound. Thus we see that the chemical shift information is augmented with fairly accurate area information regarding the number of equivalent protons.

Integrals are commonly obtained on routine spectrometers by use of a capacitor integrator that charges during the time the spectrometer sweeps through a peak and is manually discharged at the end of the peak region by the operator. This leads to the wavy line integrals sometimes shown whose height is proportional to the integral.

Such integrals are not particularly reliable under routine conditions and are extremely hard to interpret in reduced spectra in textbooks and journals. However, with the advent of cheap minicomputer and microprocessor data acquisition systems, the integrals are often calculated from the acquired data stored in a minicomputer's memory. Therefore, we will indicate the calculated values of the integrals where needed.

CHEMICAL SHIFTS AND CHARGED SPECIES

Since the chemical shift seemed to depend so much on electron density, Fraenkel and co-workers[4] decided to examine the relationship between chemical shift and charge. They prepared three common aromatic species and measured their chemical shifts. The average values for a number of solvents and counterions are shown below: in all cases, the spectra were singlets, indicating that the aromatic hydrogens were indeed all equivalent.

δ5.42 δ7.27 δ9.17

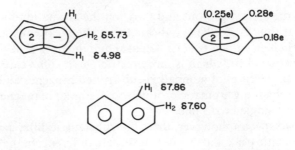

Figure 4.6 The pentalene dianion, showing its chemical shifts compared to those of naphthalene and the calculated charge densities for the pentalene dianion.

The cyclopentadienyl anion is an aromatic compound containing one negative charge and is shifted upfield from benzene by 1.85 ppm. The cycloheptatrienyl cation is an aromatic species containing one positive charge and is shifted downfield from benzene by 1.90 ppm. After allowing for the fact that a five-membered ring has a different theoretical ring current than a seven-membered ring, they arrived at adjusted shifts relative to benzene of $+2.10$ for cyclopentadienyl anion and -1.51 for tropyllium cation. Multiplying these by 5 and 7, respectively, gives approximately 10 ppm chemical shift per unit charge e, if the charge is localized on a single site.

As an example of the application of this theory, let us examine the chemical shifts of the pentalene dianion[5] shown in Figure 4.6. The spectrum consists of a triplet at $\delta 5.73$ and a doublet at $\delta 4.98$ having relative areas of $1:2$. By comparison with the model naphthalene system, an approximate charge can then be calculated for each of the carbons on pentalene. We would have expected H_1 to fall at $\delta 7.86$ and H_2 to fall at $\delta 7.60$ if uncharged. The upfield shift of each is

$$\Delta \delta H_1 = 7.86 - 4.98 = 2.88$$

and

$$\Delta \delta H_2 = 7.60 - 5.73 = 1.87.$$

We then predict a charge distribution at the carbons attached to H_1's of $0.29e$ and at H_2 of $0.19e$. This adds up to

$$4(0.29e) + 2(0.19e) = 1.54e,$$

so we conclude that the remaining charge must lie at the carbons not bearing hydrogens.

COUPLING CONSTANTS

There is much more information in nmr spectra than simple chemical shifts, however. This is apparent when you compare the spectra of 1,2-diiodoethane and iodoethane in Figure 4.7. There is only one line in the diiodoethane spectrum, since the four hydrogens are equivalent, but there are many more than two lines in the iodoethane spectrum shown in Figure 4.7b. There are, in fact, two groups of lines or *multiplets* corresponding to the two kinds of protons: methyl and methylene. These adjacent nonequivalent protons are said to be *coupled* and the distance between the lines (in Hz) in each multiplet is called a *coupling constant*. This distance is about 7 Hz in this spectrum, and we thus can write $J_{12} = 7$ Hz as the coupling constant between the methyl and methylene protons. It will be the same in both of the multiplets produced by the coupling. When nonequivalent nuclei are adjacent in a molecule, they are nearly always coupled, and the splitting induced between them is a measure of the efficiency of this interaction. We thus have a very powerful tool for determining not only what kinds of groups a molecule contains (from their shifts) but which of these groups are adjacent to which other groups (by their coupling constants). We find that the kind of splitting and the intensity of the lines in each multiplet are totally predictable from simple nmr theory.

The splitting of nmr signals into multiplets is caused by the influence of small magnetic fields from adjacent nuclear spins in the molecule. Since these fields are generated by the nuclei themselves, the magnitude of the coupling constant does not vary with the applied field. At any given instant, the nuclear spins in a molecule might well have any configuration of excited and nonexcited states, but over time there will be a certain number in each possible state. There will of course be more nuclei in states that are more probable.

Let us consider ourselves to be one of the CH_3 protons in iodoethane. The other two methyl protons are exactly like us and do not influence us in any measurable way. However, we see two CH_2 protons nearby that might have the configurations ↓↓, ↓↑, ↑↓, or ↑↑, where we use ↓ to represent the lower energy state and ↑ to represent the upper state. This is shown below:

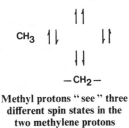

Methyl protons " see " three
different spin states in the
two methylene protons

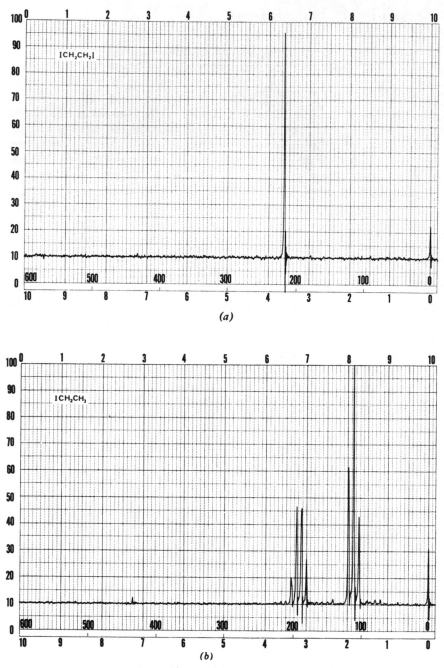

Figure 4.7 (*a*) 1,2-Diiodoethane; (*b*) iodoethane.

72

Since the methyl protons "see" a slightly diminished magnetic field when both spins are down, a slightly increased field when both spins are up, or the unmodified field when one spin is up and one down, we find that these spins influence the overall field by adding to or subtracting from it and that the absorption for the methyl group is split into three lines. Furthermore, since the middle magnetic state is twice as probable because there are two cases where there is one spin up and one down, we find that the intensities of these lines have the ratio of 1:2:1.

The magnetic field that influences the methylene protons has the same sort of additions and subtractions, and we find that there are four possible combinations of the three spins of the adjacent methyl group as shown below:

$$
\begin{array}{c}
\uparrow\uparrow\uparrow \\
\uparrow\uparrow\downarrow \quad \uparrow\downarrow\uparrow \quad \downarrow\uparrow\uparrow \\
-\mathrm{CH_2}- \quad \uparrow\downarrow\downarrow \quad \downarrow\uparrow\downarrow \quad \downarrow\downarrow\uparrow \\
\downarrow\downarrow\downarrow \\
-\mathrm{CH_3}
\end{array}
$$

The methylene protons therefore are influenced by four different fields induced by these adjacent spins, two below and two above the nominal resonance frequency. We thus find that the CH_2 protons are split into a four-line *quartet*, which has the relative intensities of 1:3:3:1.

The overall integration of the triplet and quartet as multiplets will show us that the signals have a relative area of 3:2 as predicted by the number of methyl and methylene protons, but the areas of the small peaks within these multiplets is given by the ratios shown above.

We find that, in general, the number of splittings of a multiplet adjacent to n equivalent spins is given by

$$ s = 2nI + 1 $$

where s is the number of lines and I is the spin of the nucleus causing the splitting. Since ^1H has a spin of $\frac{1}{2}$, this reduces to $n + 1$ for proton spectra, and we can remember the rule of thumb that there will be one more line than the number of equivalent protons.

The intensities of the multiplets also have a predictable ratio and turn out to be related to the coefficients of the binomial expansion $(a + b)^n$. These are given by Pascal's triangle, where each coefficient is the sum of the two terms diagonally above it:

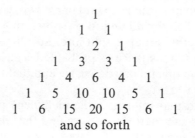

and so forth

For simple spectra, then, we can predict the number of splittings and their intensities from the multiplicity and intensity rules given above. These rules only apply when the chemical shifts are much greater than the coupling constants. Such spectra are termed *first order*. When the shifts become smaller, more complex spectra, termed *second order* spectra, result.

However, we have not yet defined the meaning of equivalent nuclei. Two nuclei are said to be *magnetically equivalent* if they have the same chemical shift and the same coupling constant to every other nucleus in the molecule. This is often the same as chemical equivalence, but there may be some subtle differences. For example, consider the following cases:

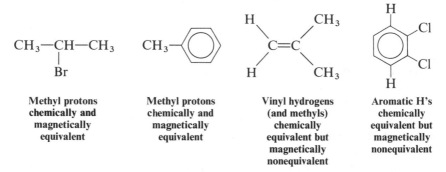

| Methyl protons chemically and magnetically equivalent | Methyl protons chemically and magnetically equivalent | Vinyl hydrogens (and methyls) chemically equivalent but magnetically nonequivalent | Aromatic H's chemically equivalent but magnetically nonequivalent |

In the case of the 2-methylpropene, one methyl group would be expected to show a different coupling to the vinyl hydrogen which is *cis* to it than to the one which is *trans* to it. Likewise, the other methyl group will also exhibit this dissimilarity, but for the opposite hydrogens. Thus both methyl groups do not have the same coupling constants to the same hydrogens and are therefore magnetically nonequivalent. Similarly, in the case of o-dichlorobenzene, the four hydrogens on the aromatic ring are not equivalent, nor are pairs of them equivalent, because H_3 will have a different coupling to H_4 than H_6 will, for example. These magnetic nonequivalences can lead to some extremely complex spectra as we will see later.

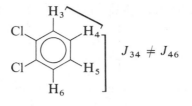

$$J_{34} \neq J_{46}$$

NAMING SPIN SYSTEMS

It is useful, on occasion, to refer to the type and number of equivalent nuclear spins in a molecule by a general name, indicative of the relation between nuclei. We do this using letters of the alphabet to indicate the similarity or difference of the nuclei involved by chemical shift. We name the first groups with letters near the beginning of the alphabet (A, B, C) and those which are markedly different by letters near the end of the alphabet (X, Y, Z). If there are any of intermediate shift, we name them by letters from the middle (M, N, O). For example, we might consider ethyl bromide to be an A_3X_2 system, where the methyl hydrogens are called "A" and those on the carbon containing the bromine called "X." In the same manner we could name 1-chloropropene as A_3XY. We might be tempted to name the aromatic hydrogens of p-bromotoluene as A_2B_2 but this colloquial usage is actually wrong since the two hydrogens *ortho* to the methyl group are not exactly equivalent.

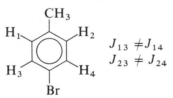

$$J_{13} \neq J_{14}$$
$$J_{23} \neq J_{24}$$

In fact, the system should be called an $AA'BB'$ system, to indicate that while the protons are chemically equivalent, they are magnetically non-equivalent. Similarly, we name 1,1-difluoroethylene as an $AA'XX'$ system (fluorine also has spin of $\frac{1}{2}$).

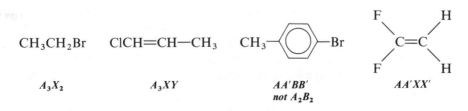

CH_3CH_2Br $ClCH{=}CH{-}CH_3$

A_3X_2 A_3XY $AA'BB'$ $AA'XX'$
                                             not A_2B_2

NONEQUIVALENCE ADJACENT TO CHIRAL CENTERS

An interesting and rather subtle case of magnetic nonequivalence occurs when two protons are attached to a carbon α to an asymmetric center such as the CH_2 group in 2-chlorobutane. If we consider the S isomer shown below in its three rotamers, we can make a table of the positions that H_a and H_b occupy. If we consider only their immediate environments,

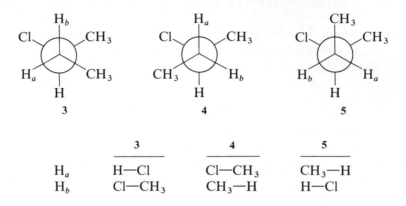

	3	4	5
H_a	H—Cl	Cl—CH_3	CH_3—H
H_b	Cl—CH_3	CH_3—H	H—Cl

it would appear that they are indeed equivalent in their environments since they cyclicly occupy the same positions between the H and Cl, the Cl and CH_3, and the CH_3 and H. However, if we consider the secondary environments we will see that when H_a is in position **3** the CH_3 on the front carbon is between the CH_3 and the H of the back carbon, while when H_b occupies this same position in **5** the CH_3 on the front carbon is between the Cl and the CH_3 on the back carbon. This is shown below:

	3	4	5
	CH_3	CH_3	CH_3
H_a	H—Cl (CH_3—H)	Cl—CH_3 (H—Cl)	CH_3—H (Cl—CH_3)
H_b	Cl—CH_3 (CH_3—H)	CH_3—H (H—Cl)	H—Cl (Cl—CH_3)

Considering the above information we see that while the hydrogens H_a and H_b occupy all the same rotational positions, the relative position of the CH_3 at the same time is different. Thus H_a and H_b are nonequivalent because their secondary environments are different. We often find rather different chemical shifts for the two protons on a carbon α to a chiral center and must treat them as *nonequivalent* in analyzing an nmr spectrum. Such protons are sometimes termed *diastereotopic*.

CHEMICAL SHIFTS IN CHIRAL MOLECULES

Protons attached to chiral centers have identical chemical and magnetic properties in achiral solvents, as we would expect. However, when these compounds are dissolved in chiral solvents, the magnetic properties will vary, because of the differing solvent environments of the protons in right- and left-handed molecules. Thus a racemic mixture will give only a single peak for a proton attached to a chiral center when the solvent is achiral, but may resolve into separate resonances when the mixture is dissolved in a chiral solvent.

When two protons are attached to a carbon bearing two other differing groups, these protons are termed *enantiotopic*. If one of the protons is replaced with a fourth different group, a chiral center would be formed. If the other proton is replaced instead with this fourth group, the enantiomer of the first chiral compound would be formed. While enantiotopic protons are equivalent in both the chemical and magnetic sense in achiral solvents, they become magnetically nonequivalent in chiral solvents and thus may give rise to different resonances. For example, the protons on C-2 of butane may be termed enantiotopic, since replacing either of them with any group that is not CH_3, CH_3CH_2 or H will cause a chiral center to be formed.

INTERPRETATION OF SIMPLE NMR SPECTRA

With the above fundamentals in mind, let us now look at a few nmr spectra. Figure 4.8 shows the spectrum of 1-bromopropane. The spectrum apparently consists of two triplets and a sextet. This would not at first seem to correlate with the structure, so let us examine the couplings in detail. The methyl group is adjacent to a $—CH_2—$ group and thus would be expected to be a triplet $(2 + 1 = 3)$, and the $—CH_2Br$ group is also adjacent to a $—CH_2—$ group and is similarly a triplet. The upfield triplet at $\delta 1.3$ is due to the methyl group and the downfield one, which is deshielded by the Br, is due to the CH_2Br group.

The central sextet is at first somewhat puzzling. A cursory examination of the spacing between the peaks in the two triplets indicates that they have about the same coupling constants, around 6–7 Hz. If one multiplet is split by 6 Hz, then the other multiplet which is coupled to it must also be split by 6 Hz, and we really have a central $—CH_2—$ group split equally by both the methyl and the CH_2Br groups. To predict the splittings of the resulting multiplet, we can draw a tree graph, in which we predict one splitting and then the other. We

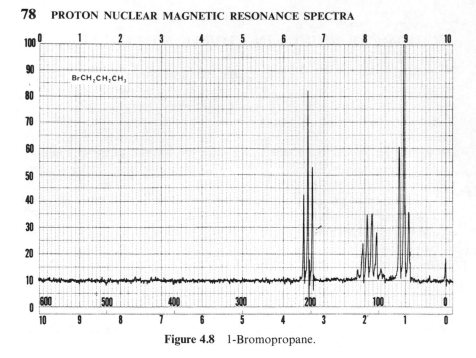

Figure 4.8 1-Bromopropane.

start by drawing a single line for the unsplit CH_2 and then indicate that it would be split into a triplet by the —CH_2Br:

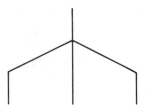

Then we split each of the resulting three lines into quartets, predicting the methyl group splittings. The spacings between the lines must be proportional to the coupling constants, which in this case are all about 6 Hz.

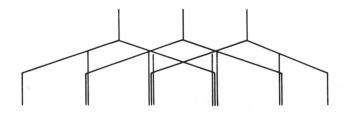

Then we find the total of the lines and predict the resulting intensities by the number of lines that come out on top of each other:

This gives a sextet whose numerical intensities can be found exactly by adding those of the quartets, remembering that the middle line of the triplet will be twice as intense to start with:

$$
\begin{array}{ccccccc}
1 & 3 & 3 & 1 & & \\
2\,(1 & 3 & 3 & 1) & & \\
& & 1 & 3 & 3 & 1 \\
\hline
1 & 5 & 10 & 10 & 5 & 1
\end{array}
$$

Thus a sextet is predicted, having intensities of $1:5:10:10:5:1$, and this is exactly what is observed in Figure 4.8. It should be noted here that the intensities and splittings are the same as would have been predicted if the central methylene group were coupled to five *equivalent* hydrogens, although they are clearly not equivalent in this compound. This simplification will only occur when (a) the coupling constants are essentially equal, and (b) when the chemical shifts are much greater in Hz than the coupling constants.

Let us now consider the spectrum in Figure 4.9, which has the molecular formula C_3H_7I. This spectrum shows a strong doublet at about $\delta 1.9$ and a multiplet at $\delta 4.3$. Without further chemical information we might initially conclude that this is a triplet or a quintet at $\delta 4.3$, indicating one hydrogen of one kind and two or four of another kind. However, accurate examination of the intensity information in this spectrum indicates that this is certainly not a triplet since the intensities are not $1:2:1$, and probably not a quintet either since the intensities are probably not $1:4:6:4:1$. If it is neither, then the integrals may be of use. In this case, they show that the ratio of the total areas is about $6:1$. This clearly indicates that we are dealing with six hydrogens of one kind and one of another. In this case, the ratios would be $1:6:15:20:15:6:1$. Since the lowest peaks are so very weak compared to the rest, it is common to see only five of the seven lines in a septet without further amplification. The compound is 2-iodopropane:

$$CH_3-\underset{\underset{I}{|}}{CH}-CH_3$$

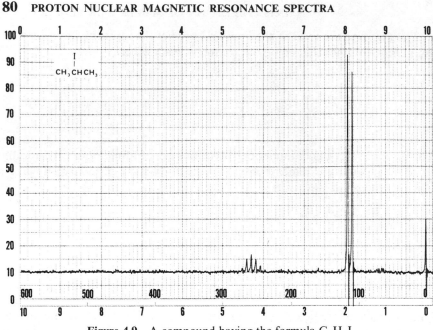

Figure 4.9   A compound having the formula C_3H_7I.

Note that the six equivalent protons in this compound are not all on the same carbon, but are on two chemically equivalent carbons. They are, nonetheless, magnetically equivalent since they all have the same coupling constant to the central hydrogen. The isopropyl hydrogen is split into a septet by these six equivalent protons, and the methyl protons are split into a doublet by the one isopropyl hydrogen. This isopropyl pattern is a very common one and easily recognized even in spectra containing many more groups.

MAGNITUDE AND SIGN OF COUPLING CONSTANTS

We have seen that there is a great deal of information available from the careful analysis of the splitting patterns in proton nmr spectra. While the coupling constants in these simple examples have been about 6–7 Hz, J-values actually vary over a rather wide range as shown in Table 4.3.

It is apparent from this table that coupling constants can vary in sign as well as in magnitude. The physical meaning of the sign of a coupling constant is related to whether the two coupled nuclei are aligned or opposed in their lowest energy state. A simplified method of illustrating this can be shown

Table 4.3 Proton Coupling Constants

Group	J_{HH}
H—C—C—H	7

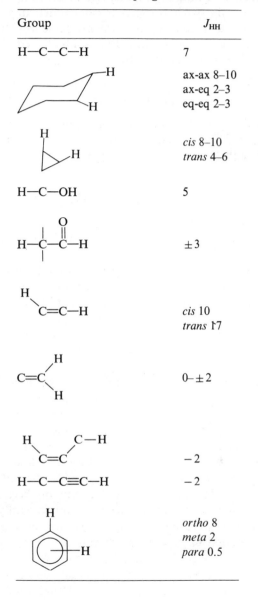

	ax-ax 8–10
	ax-eq 2–3
	eq-eq 2–3
	cis 8–10
	trans 4–6
H—C—OH	5
	±3
	cis 10
	trans 17
	0–±2
	−2
H—C—C≡C—H	−2
	ortho 8
	meta 2
	para 0.5

by the electron-mediated coupling model used in the following two illustrations. Let us consider the two hydrogens as having their spins coupled by the interaction of the electron pair bonds in the molecule. By Hund's rule all of the electrons on a single carbon will have their spins aligned. For a three-bond coupling H—C—C—H, we could write

$$\overset{\uparrow}{H}\downarrow C\uparrow\downarrow C\downarrow\uparrow\overset{\downarrow}{H}$$

and we find that if one nuclear spin H has one spin " up," then the other will be " down," thus leading to a *positive* coupling constant. On the other hand, for allylic coupling through four bonds, we could write

$$\overset{\uparrow}{H}\downarrow C\uparrow\downarrow C\downarrow\uparrow C\uparrow\downarrow\overset{\uparrow}{H}$$

and find that if one nuclear spin is "up," the other will also be "up." This is a *negative* coupling and is indeed the one generally observed in allylic systems.

High resolution nmr spectra are not dependent on the absolute sign of the coupling constants, but only on their relative sign. Thus, if we change the sign of all the couplings in an nmr spectrum, its appearance will not change. However, if we change the signs of only some coupling constants in a closely coupled spectrum, some of the line intensities will change markedly. There will be no effect on a first-order spectrum.

The absolute sign of coupling constants has only been determined in a few cases since this process involves complex experiments such as analysis of spectra obtained in liquid crystals,[6] but most signs of coupling constants are now known by relating these few to other couplings in complex spectra.

The most common coupling is over three-bonds such as H—C—C—H. However, both closer and more distant ones are known that vary according to the electronegativity of the groups, conjugation with π-systems, and the stereochemistry. For example, the coupling over four-bonds in allylic systems (H—C=C—C—H) is about -2 Hz, presumably because of π-system transmissions of spin information. We see an example of allylic coupling in the spectrum of 3-chloro-2-methylpropene shown in Figure 4.10. The spectum appears at first glance to be composed of one doublet and two singlets at $\delta5.02$, $\delta4.03$, and $\delta1.86$ due to the $=CH_2$, CH_2—Cl, and CH_3, respectively. However, closer examination reveals additional splittings in the doublet at $\delta5.02$ and the singlet at $\delta1.86$, although no further splittings are detected in the other singlet. Thus these two distant protons are coupled by a four-bond allylic coupling. Since the two vinyl protons are not equivalent, the splitting is not expected to be simple, and overlap of these fine splittings is indeed observed. Similarly, the three-bond coupling through a

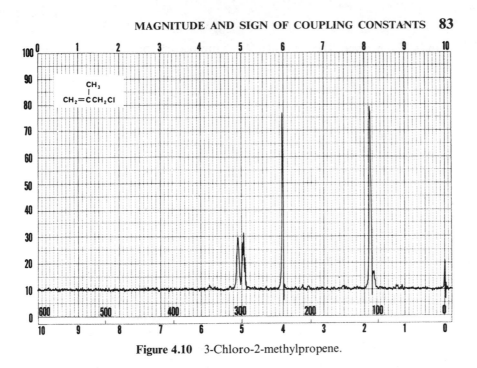

Figure 4.10   3-Chloro-2-methylpropene.

double bond is much larger than through a single bond; for example, in H—C=C—H, J is 10–17 Hz compared to only 7 Hz in the saturated system.

Benzylic coupling constants are also fairly large, considering that they occur across as many as six bonds. The *ortho*, *meta*, and *para* couplings between ring hydrogens and the methyl hydrogens in toluene have been measured at 0.6, 0.35, and 0.5 Hz, respectively.

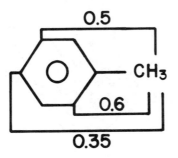

Long-range couplings are also observed between protons at the ends of a conjugated chain held in a zig-zag configuration, and in saturated four-bond systems held rigidly in a conformation shaped like an M or W. This latter case is sometimes called the Cassiopeia effect and is observed in compounds like **6** where the requisite W shape occurs, but not in **7** where it does not.

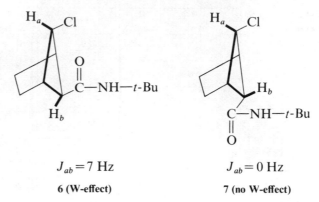

$J_{ab} = 7$ Hz $\qquad\qquad J_{ab} = 0$ Hz

6 (W-effect) $\qquad\qquad$ **7 (no W-effect)**

VARIATION OF COUPLING CONSTANTS WITH ANGLE

By valence bond theory, Karplus[7] worked out a dependence of the vicinal coupling $J_{H-C-C-H}$ and the dihedral angle of the fragment. It may be written approximately as:

$$J_{HCCH} = 7.0 - \cos\theta + 5\cos 2\theta$$

This is plotted in Figure 4.11. The most important conclusion of this theory, which has been amply verified by experiment, is that the coupling constant

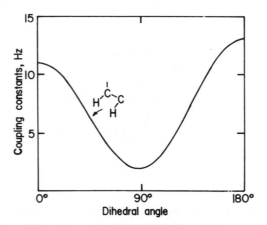

Figure 4.11 A plot of the dependence of vicinal coupling $J_{H-C-C-H}$ versus dihedral angle as reported by Karplus.

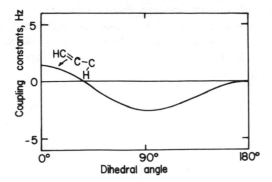

Figure 4.12   A plot of the dependence of allylic coupling $J_{H-C=C-C-H}$ versus dihedral angle.

reaches a maximum for a dihedral angle of 180°, a lesser maximum for $\theta = 0°$, and that the coupling constant is approximately 0 for a dihedral angle of 90°. This theory and its experimental verification have great implications in conformational analysis of semirigid systems, since it has been found that J_{aa} for cyclohexanes is about 10 Hz and J_{ee} and J_{ae} are only 2–3 Hz as predicted from their dihedral angles.

A similar relationship has been developed for allylic couplings[8] giving a curve of roughly the same shape, except that it passes through zero and becomes negative, making the dihedral angle of 90° the one having the largest (negative) coupling constant. For

$$H_a-C=C-C-H_b$$

J_{ab} is given by Figure 4.12. The electronegativity of substituents affects the size of these coupling constants, so these graphs must be used with caution.

GEMINAL COUPLING CONSTANTS

Very often spectroscopists make reference to the coupling constant in molecules like methane or ethylene as -12.6 or $+2.3$ Hz. Clearly these have not been observed in conventional nmr spectra since the protons in question would be expected to produce only a singlet. The protons are *coupled*, however, but since they are magnetically equivalent, the spectral lines are not *split*. We can observe this coupling by substituting a deuterium for one of the protons and measuring J_{HD}. Then J_{HH} is obtained from the magnetogyric ratios of hydrogen and deuterium:

$$J_{HH} = \frac{\gamma D}{\gamma H} J_{HD} = 6.51 \, J_{HD}$$

In sp^2 hybridized carbons, the coupling constant between two hydrogens on the same carbon is small (0–2 Hz), but becomes larger when electron donating groups are attached to the carbon. Thus the coupling constant in formaldehyde is +42 Hz and +17 Hz for *N-tert*-butylformamide. Similarly, electron withdrawing groups

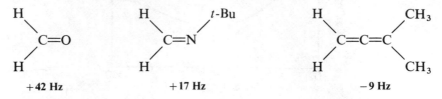

decrease the efficiency of coupling and actually make the coupling constant negative. For example, the coupling constant in 1,1-dimethylallene is −9 Hz. For many terminal methylene groups not attached to atoms of markedly different electronegativity, the coupling constant is near zero.

On sp^3 carbons, the coupling constants vary from −12.4 for methane to −20.4 for $NC—CH_2—CN$. Again, more electron withdrawing groups make J_{HCH} more negative and electron releasing groups make it more positive. These sp^3 couplings are not actually observed in the unmodified compounds, but only in rigid systems such as small rings where the introduction of one or

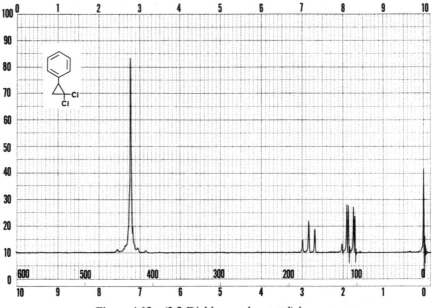

Figure 4.13 (2,2-Dichlorocyclopropyl)-benzene.

more substituents makes the two nuclei nonequivalent. For example, in Figure 4.13 the spectrum of (2,2-dichlorocyclopropyl)-benzene becomes an *ABX* system where the two geminal cyclopropyl protons are nonequivalent.

COUPLINGS WITH OTHER NUCLEI

As we saw earlier, ^{1}H is not the only nucleus with a nuclear spin, and couplings between ^{1}H and these nuclei are quite common. The nuclei having substantial abundance and nonzero spin that occur most frequently in organic compounds are ^{13}C, ^{19}F, and ^{31}P. All of these have a spin of $\frac{1}{2}$. While we will deal with ^{13}C spectroscopy separately in Chapter 6, we should point out here that small satellite peaks 60–125 Hz on either side of a strong peak are often the splittings due to $J_{^{13}C-H}$. These vary in size with hybridization as shown below.

Hybridization	$J_{^{13}C-H}$ (Hz)
sp^3	125
sp^2	170
sp	250

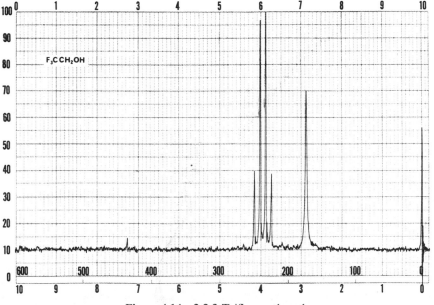

Figure 4.14 2,2,2-Trifluoroethanol.

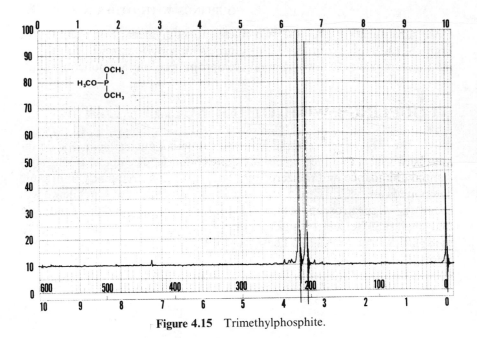

Figure 4.15 Trimethylphosphite.

Table 4.4 Proton-Heteroatom Coupling Constants

Group	J_{HX}
H—C—F	45
H—C—C—F	5–20
H—C—C—C—F	5
H—C=C—F	*cis* 18 *trans* 40
$\begin{array}{c}\diagdown\\ \diagup\end{array}C=C\begin{array}{c}{}^H\\ {}_F\end{array}$	80
H—P$\diagup$	200
HĊ—P=O	10

These must be carefully distinguished from spinning sidebands introduced by tube wobbling which will vary in frequency depending on the spinning rate.

Coupling constants between ^1H and ^{19}F can be large, even through several bonds. In the spectrum of 2,2,2-trifluoroethanol shown in Figure 4.14, we see a singlet due to the OH and a quartet due to the CH_2, which is split by the three adjacent fluorines.

Large coupling constants are also observed between ^{31}P and ^1H. The spectrum of trimethyl phosphite shown in Figure 4.15 shows J_{31P-H} of nearly 12 Hz through three bonds. Some of these coupling constants are tabulated in Table 4.4.

"THROUGH SPACE" OR ELECTRON PAIR MEDIATED COUPLING

In a few significant cases, spin–spin coupling appears to occur without the influence of intervening bonds. In these cases, the mechanism appears to involve an unshared electron pair that transmits the spin information through space to a nucleus held rigidly nearby. For example, the coupling between H_1 and H_2 in compound **8** shown below is 1.1 Hz.[9] Since the lone pair on the oxygen is held rigidly near the H_2, this mechanism seems more reasonable than a six-bond coupling through a saturated chain.

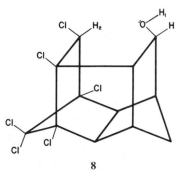

8

The five bond F—F coupling in a number of substituted difluoro-methylanthracenes (**9**) is uniformly in the range of 167–178 Hz regardless of

9

the nature of X. Since electron withdrawing groups might be expected to affect any through-bond coupling mechanism, the electron pair mediated mechanism through space is preferred as an explanation here as well.[10]

SECOND-ORDER SPECTRA

Thus far we have looked only at fairly simple nmr spectra, whose chemical shifts and coupling constants can be deduced more or less by inspection. However, when we look at the spectrum of 3-buten-2-ol shown in Figure 4.16, we see a veritable forest of lines (more than we might at first think possible). Clearly the doublet at $\delta 1.12$ is caused by the methyl group and the singlet is probably caused by the OH. Furthermore, the broadened quintet is most probably due to the CH—OH hydrogen, coupled both to the methyl and the vinyl CH protons. However, there are 12 other lines in the spectrum where we might previously have only predicted 5: a triplet and a doublet. To understand this, let us consider a three-spin system made up solely of the $=CH_2$ and the $=CH$. We find that there are eight possible spin states:

$$\uparrow\uparrow\uparrow$$
$$\downarrow\uparrow\uparrow \quad \uparrow\downarrow\uparrow \quad \uparrow\uparrow\downarrow$$
$$\downarrow\downarrow\uparrow \quad \downarrow\uparrow\downarrow \quad \uparrow\downarrow\downarrow$$
$$\downarrow\downarrow\downarrow$$

There are indeed 15 possible transitions among these eight levels in which there is only a net spin change of 1. One of these is "forbidden" and two others are very weak. Thus, while we might expect 15 lines in all such cases, actually we only observe all these transitions as having *different* energies when the chemical shifts are not large compared to the coupling constants. The system made up of the $=CH_2$ and the $=CH$ is an *ABX* system in which the two $=CH_2$ protons have very similar chemical shifts and the $=CH$ proton has a somewhat different shift. In fact, if we inspect Figure 4.16 very closely, we find that the region contains even more than 15 lines because the entire *ABX* pattern is further split by the CH—OH proton, and the spin system comprising this region might better be referred to as *ABMX* to include all four interacting protons. In these cases the spectra become *second order* and *neither the chemical shifts nor the coupling constants* can be obtained by direct inspection. Systems containing this *ABX* pattern are so common in nmr that we will discuss how to analyze them in the next section. For the moment let us remember that many nmr spectra, in fact most spectra of "interesting" compounds, have some second-order features that cannot be analyzed directly. Another of these highly complex second-order spectra

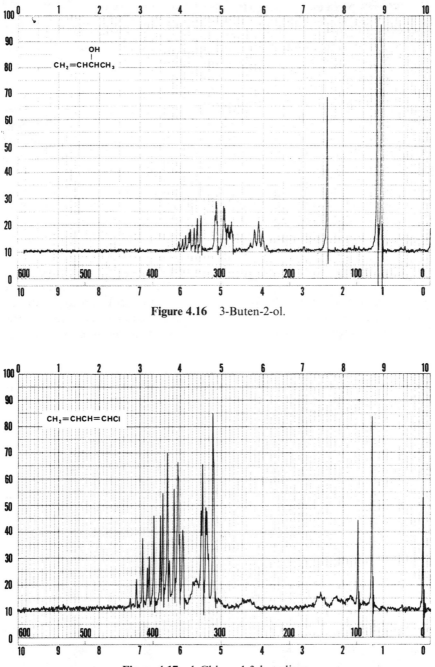

Figure 4.16 3-Buten-2-ol.

Figure 4.17 1-Chloro-1,3-butadiene.

is shown in Figure 4.17 for 1-chloro-1,3-butadiene, where again, the complex couplings come from spin-coupling information transmitted by the π-system. In such systems, tree-diagrams cannot be used!

Other common examples of complex second-order spectra occur in such systems as aromatic rings with a highly electronegative substituent, such as bromine or fluorine. Figure 4.18 shows the spectrum of bromobenzene, which might be referred to as an *ABB'MM'* system, where the two protons α to the bromine are *M* and *M'* and the three most distant are *A*, *B*, and *B'*. This system is again quite complex, since the coupling constants range from 0.5–8 Hz and the shifts are spread over only 0.5 ppm or about 30 Hz at 60 MHz.

To be able to understand what happens in second-order spectra and to recognize the characteristics of these spectra, let us consider a simple *AX* system. We will slowly decrease the chemical shift between the nuclei *A* and *X* until it becomes *AM*, *AB*, and finally *AA'*, as simulated in Figure 4.19. As we see, the system starts out as two doublets with nearly equal intensity in all four lines, and as the shift between them decreases, the intensity of the outer lines begins to decrease. We say that the doublets begin to " lean in " toward the center when we observe this, which is the first indication of a non first-order spectrum. Recall that a simple first-order analysis would have

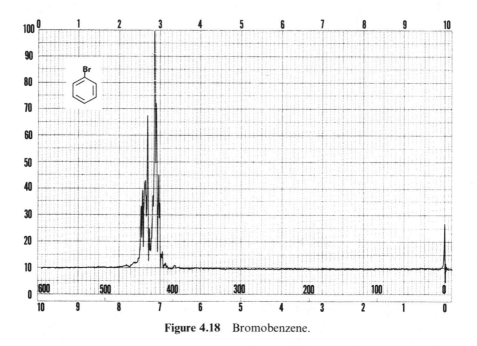

Figure 4.18 Bromobenzene.

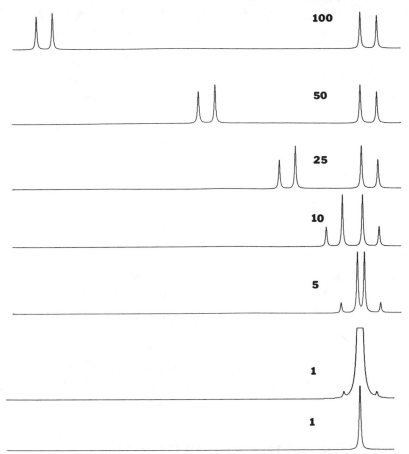

Figure 4.19 Theoretical calculation of two-spin system showing how the intensities and spacings vary as the system changes from AX (top) to AM (2nd, 3rd) to AB (4th, 5th) to AA' (bottom two traces). In all cases, the coupling constant $J_{ab} = 5$ Hz. The chemical shift in Hz is shown on the diagram. The bottom trace shows a chemical shift difference of only 1 Hz and is expanded in the trace above it to show the very small outer lines.

predicted equal intensities for all four lines. Finally as AM becomes AB, the outer lines begin to disappear and, as we expect, nearly a singlet is observed in the AA' case.

We often see the beginnings of second-order effects in spectra that can almost be analyzed as first order, such as in the spectrum of ethyl sulfide shown in Figure 4.20, where we see the triplet and quartet leaning in toward

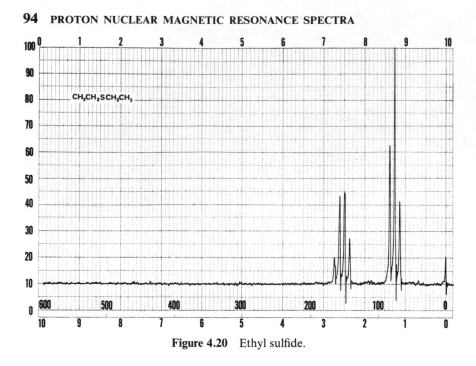

Figure 4.20 Ethyl sulfide.

each other. In these cases, the exact chemical shift may not be at the center of the multiplet as expected and careful analysis is advised if exact shifts are desired.

ANALYSIS OF THE AB SPECTRUM

When the spectrum is first order, in the AX case, we can confidently measure the coupling constant as the distance in Hz between the lines in either doublet and the chemical shift as the distance between the center of the two doublets. However, as the system becomes more like AB, we find that while the coupling constant is still the distance between the lines of the doublet, the chemical shift is *not* the distance between the center of the multiplets. For example, when the AX doublets are 80 Hz apart, we measure the frequencies of the four lines in Hz as:

1	7.42
2	12.42
3	87.58
4	92.58

The coupling constant is clearly 5.00 Hz, since it is the difference between lines 1 and 2 or 3 and 4. The chemical shift between the two would be predicted by simple first-order analysis to be the distance between the midpoints of the two doublets or

$$\frac{(92.58 + 87.58)}{2} - \frac{(12.42 + 7.42)}{2} = 80.16 \text{ Hz.}$$

This is quite close to the actual value of 80.00 Hz and would lead to no substantial error.

However, if we examine the spectrum in which the peaks are separated by only 5 Hz, we find that the peaks are:

1	6.46
2	11.46
3	13.54
4	18.54

and our simple calculation would yield:

$$\frac{(18.54 + 13.54)}{2} - \frac{(11.46 + 6.46)}{2} = 7.08$$

which is a 29% error from the actual value of the chemical shift of 5.00 Hz.

The correct formula that can be applied ın any case from AX to AB is:

$$(v_1 - v_2) = [(F_1 - F_4)(F_2 - F_3)]^{1/2}$$

When applied to the values given above this yields 5.001 Hz as predicted.

The intensities of the lines can also be predicted from their separation by the following formula:

$$\frac{I_2}{I_1} = \frac{I_3}{I_4} = \frac{(F_1 - F_4)}{(F_2 - F_3)}$$

One interesting implication of this formula is that if only one of the doublets is visible, the chemical shift of the other can be calculated from the intensities.

THE ABX SYSTEM

The ABX system is common enough to warrant special discussion. It occurs in many $CH_2{=}CHR$ vinyl systems and in 1,2,4-substituted benzenes among others. We often need to know the exact chemical shifts and coupling constants in an ABX system in order to compare them with those in other model systems. From these parameters, we can often infer information about conformation, angle strain, and hybridization.

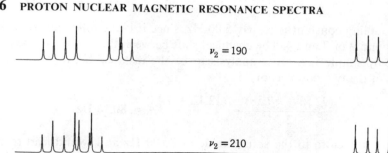

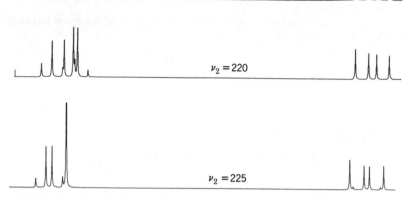

Figure 4.21 Theoretical calculation of a three-spin system showing the variation in line positions and intensities as shifts vary from AMX to ABX. $v_1 = 20$, $v_3 = 230$, and v_2 varies from 190–220. $J_{12} = 8$, $J_{13} = 15$, and $J_{23} = 7$.

To appreciate how an ABX system develops and changes with various shifts, let us consider the AMX to ABX system transition simulated in Figure 4.21. We see at first the three doublet-of-doublets multiplets predicted for three widely separated spins coupled to each other. Then as the central multiplet is moved downfield so that AMX approaches ABX, we see a change in the intensities of the X part and a substantial change in the AB (left-hand) part until the shifts and couplings become extremely complex indeed. As the X-part moves toward the AB part, the spectrum becomes ABC, more lines appear, all 15 transitions actually occur, and the spectrum can only be analyzed with the help of computer programs. A thorough discussion of hand analysis of the ABX system has been given by Becker.[11]

DECEPTIVELY SIMPLE SPECTRA AND "VIRTUAL COUPLING"

The ABX system leads to some unusual looking spectra that have been given the name "deceptively simple spectra." These occur when some of the lines in the pattern coincide. In Figure 4.22 we see an $AA'X$ spectrum in which the

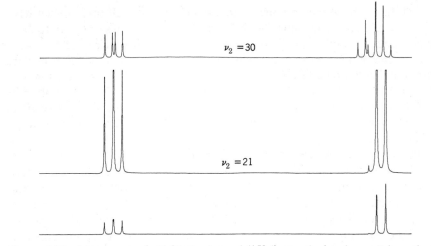

Figure 4.22 Movement of *AMX* (top) to *AA′X* (bottom) showing a "deceptively simple" spectrum. $v_1 = 20$ and $v_2 = 30$ (top), 21 (bottom), and $v_3 = 200$. $J_{12} = 5$, $J_{23} = 7$, and $J_{13} = 5$. The spectrum on amplification (middle trace) shows additional lines of the *ABX* pattern, which are often lost in the noise of the real spectrum.

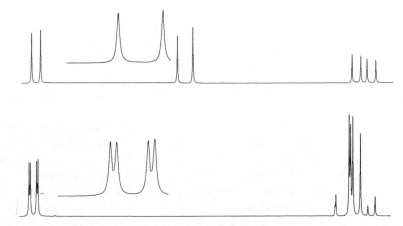

Figure 4.23 Variation of *AMX* to *ABX* where $J_{ab} = 10$, $J_{ac} = 6$, and $J_{bc} = 0$. As *AMX* approaches *ABX* the spectrum ceases to be first order and splittings appear in the downfield peak as if J_{bc} were nonzero. This effect is often termed "virtual coupling," but is in reality a special case of a second-order spectrum that cannot be analyzed simply.

97

two shifts in the A part are nearly identical, although all three nuclei have nonzero coupling constants. Without amplification, these appear to be a triplet and a doublet, but with sufficient amplification some of the other lines in the pattern appear. In such a spectrum, the observed splittings are not actual coupling constants, but are the *average* of coupling constants, in this case of J_{AX} and $J_{A'X}$. These accidental coincidences of lines can often be resolved by obtaining the spectrum in a different solvent, which may affect the chemical shifts slightly but not the coupling constants, or by using a different field strength.

In Figure 4.23, we see an AMX spectrum where $J_{12} = 10$ and $J_{13} = 6$, but $J_{23} = 0$. Since v_3 is coupled only to v_1, we would expect that the pattern of v_3 would be merely a doublet. In the ABX case, however, where v_1 and v_2 are separated by only 10 Hz, we see that additional splittings have developed in v_3. This phenomenon was originally named "virtual coupling" since the splittings occurred even though there was no coupling constant between v_2 and v_3. It is more accurate, however, to regard this simply as the consequence of a non-first-order spectrum, in which we do not *expect* simple splitting patterns to occur.

SPIN DECOUPLING TECHNIQUES

It is quite possible to analyze spectra of up to seven or eight spins by computer methods or, in the simpler cases, by hand. Most of these techniques require at least some initial guessed parameters to iterate for a better fit. However, when your purpose in analysis of the spectrum is compound identification, guessing may not be productive. It is possible in this case to obtain more information about the spectrum by *decoupling* some of the spins from each other.

Decoupling is accomplished by applying a second strong field H_2 to some line or multiplet that is coupled to another and examining the resulting spectrum. When a multiplet is irradiated strongly, it becomes *saturated*, and flips back and forth between its two states so rapidly that it no longer affects any adjacent nuclei. When this happens, the spectrum simplifies, since the coupling between that nucleus or group of nuclei and all others has been removed. Often, extremely complex spectra can be reduced to first-order cases by using spin decoupling to remove some of the interactions.

Figure 4.24 shows the spectrum of *trans*-crotonaldehyde obtained as a single resonance spectrum, upon irradiation of the methyl group, and on irradiation of the aldehyde proton. Since allylic coupling makes this an A_3XYZ spectrum, it is nearly impossible to analyze. However, when the methyl group is decoupled from the three olefinic protons, the spectrum

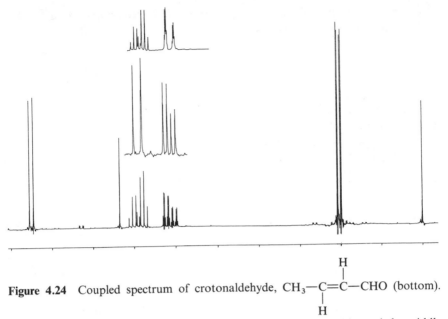

Figure 4.24 Coupled spectrum of crotonaldehyde, CH_3—C=C—CHO (bottom).

The upper trace shows the result of irradiating the aldehyde doublet and the middle trace shows the result of irradiating the methyl multiplet.

becomes an *ABX* (or *ABC*) case which can be analyzed much more readily. In modern nmr terminology, we refer to this technique as *homonuclear* decoupling, since we are decoupling nuclei of the same kind from each other. This is distinguished from *heteronuclear* decoupling where different kinds of nuclei, such as ^{13}C and ^1H, are decoupled.

RATE PROCESSES OBSERVED BY NMR

In IR spectroscopy, we discovered that if two species such as two conformers of a molecule were present, we could observe both of them in the spectrum. The same generalization will be found to be true in UV spectroscopy. However, in nmr the observation frequencies are much lower, and substantial precession at a given Larmor frequency may not have occurred before one species interconverts to another.

According to the Heisenberg uncertainty principle,

$$\Delta E \cdot \Delta t \cong h$$

where ΔE is the uncertainty in energy corresponding to an uncertainty in the time of measurement Δt. Now since $\Delta E = 2\pi h \Delta v$, we can write

$$\Delta t = \frac{h}{\Delta E} = \frac{1}{2\pi \Delta v}$$

where Δv is the change in chemical shift between species and Δt the mean lifetime of the species.

Now suppose we consider first an IR spectrum in which lines corresponding to two interconverting species are 6 cm^{-1} apart. Recall that

$$v \text{ in cm}^{-1} = c(v' \text{ in Hz}), \tag{1.3}$$

if we want to calculate the lifetime corresponding to coalescence of these lines, we have

$$\Delta t = -\frac{1}{2\pi(6 \text{ cm}^{-1})(3 \times 10^{10} \text{ cm/sec})} = 8.8 \times 10^{-13} \text{ sec.}$$

Similarly in UV, if an absorption is found at 250 nm and another at 260 nm the lifetimes such that the peaks would coalesce are given by

$$\Delta t = \frac{\Delta \lambda}{2\pi c} = \frac{10 \times 10^{-7} \text{ cm}}{2\pi(3 \times 10^{10} \text{ cm/sec})} = 5.3 \times 10^{-18} \text{ sec.}$$

However, in nmr, the observing frequency is only about 100 MHz and if the lines are 50 Hz apart

$$\Delta t = \frac{1}{2\pi(50 \text{ sec}^{-1})} = 3 \times 10^{-3} \text{ sec,}$$

a reasonable lifetime for a chemical species.

We can thus observe interconverting species as individual lines, broad lines, or a sharp singlet depending on the chemical lifetime of the species. For example, cyclohexane is a singlet at room temperature, even though there are both axial and equatorial hydrogens present, each having different chemical shifts. The rapid flipping of the cyclohexane molecule from one chair form to the other causes all of the protons to become equivalent over time, and we see a single sharp line at the average chemical shift for the different kinds of protons. At lower temperatures, however, the spectrum becomes much more complex.

One of the examples of this phenomenon most often observed occurs in amides whose rotation is somewhat restricted by conjugative interaction

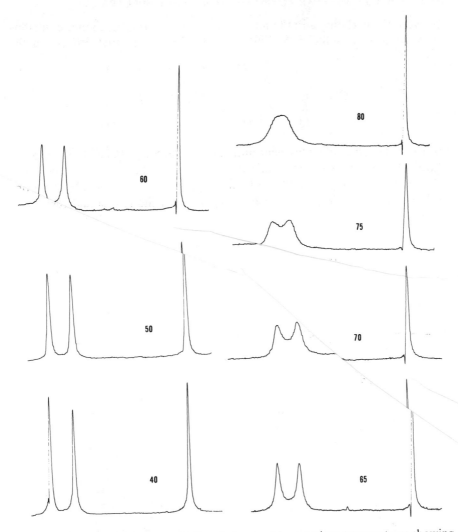

Figure 4.25 The spectra of *N,N*-dimethylacetamide at various temperatures showing the barrier to rotation at low temperatures, where the two methyl groups are non-equivalent. At higher temperatures the rotation rate increases and a single peak at the average chemical shift is shown.

between the carbonyl and the nitrogen. Because of this resonance contribution, rotation is sufficiently hindered that the two methyl groups on the

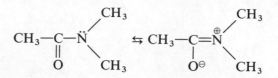

nitrogen become nonequivalent and show separate peaks at low temperatures. As the temperature is raised, the two lines begin to coalesce, as is shown in Figure 4.25. Finally, at high temperatures, only a single line is observed, having a chemical shift midway between the two actual resonance lines.

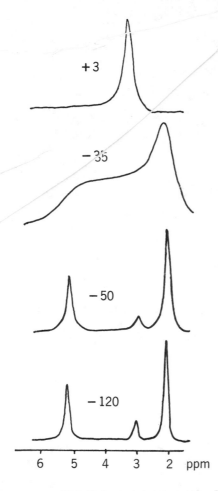

+3

−35

−50

−120

6   5   4   3   2   ppm

Figure 4.26 The spectra of the carbocation produced by treating 2-*exo*-norbornyl fluoride with "magic acid" (SbF_5—SO_2—SO_2F_2) at +3°, −35°, −50°, and −120°. (Reproduced by permission of the American Chemical Society.)

A second related rate phenomenon that may be observed in nmr is chemical exchange in which a proton actually migrates from one site to another. The most common example of this is hydrogen bonding in alcohols, where we observe that most alcohol OH peaks are broad singlets rather than multiplets coupled to the adjacent protons. This exchange can be slowed and the coupling observed in DMSO solution where hydrogen bonding to the solvent replaces any intramolecular hydrogen bonding and consequent exchange. Furthermore, scrupulously pure alcohols having no traces of water or acid will also show the coupling of the OH protons.

One of the more interesting cases of chemical equilibria that has been studied by nmr is the behavior of the norbornyl cation at various temperatures in a landmark paper by Saunders, Schleyer, and Olah.[12] At room temperature this compound, prepared by treating 2-*exo*-norbornyl fluoride with SbF_5—SO_2—SO_2F_2 solution, shows a single sharp line at about $\delta 3.2$ as shown in Figure 4.26. As this solution is cooled the line broadens to show two components, and at $-50°$ and below three components are found, having relative intensities of 4:1:6. The interpretation of these spectra postulates three conversion processes:

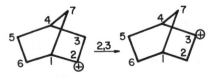

Wagner-Meerwein rearrangement

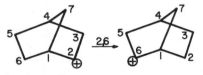

2,6-hydride shift

2,3-hydride shift

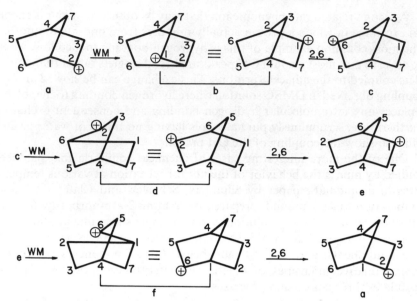

Figure 4.27   Low-temperature interconversion processes for the norbornyl action.

At low temperatures, there are three different kinds of protons having the ratio of 4:1:6, which means that only the first two of these mechanisms contribute as shown in Figure 4.27. Thus protons on carbons 1, 2, and 6 interconvert giving the peak of area 4. Note that these protons are shifted very far downfield (δ5.2) since they are alternately attached to a carbon bearing .

interconversion of 1, 2 and 6				interconversion of 3, 5 and 7		
	1	2	6		3	5	7
a	b	$\oplus$	β	a	α	γ	1
b	$\oplus$	b	β	b	1	γ	α
c	β	b	$\oplus$	c	1	α	γ
d	β	$\oplus$	b	d	α	1	γ
e	$\oplus$	β	b	e	γ	1	α
f	b	β	$\oplus$	f	γ	α	1

2 each of
3 possible positions
4 equivalent H's
b = bridgehead

2 each of
3 possible positions
6 equivalent H's

the positive charge. The protons on carbons, 3, 5, and 7 also interconvert giving the peak of area 6, and the single proton at bridgehead 4 gives the remaining peak of area 1. As the temperature is raised, the 2, 3 hydride shift begins to contribute and 2 = 3, 5 = 6, and 1 = 7. The proton at bridgehead 4 remains unchanged, but has a chemical shift about the same as the average shift for the 1, 2, 6 and the 3, 5, 7 peaks, so it is buried under the averaged peak shown at high temperature.

NMR SPECTRA AT HIGHER FIELDS

When a complex nmr spectrum has a forest of overlapping lines, it is sometimes useful to obtain the spectrum at a higher magnetic field. Since the chemical shifts are a function of magnetic field, spectra at higher fields will cause the chemical shifts to spread out over a wider frequency range. However, the coupling constants do not spread out in the same manner since they are a function of nuclear magnetic interactions and are not influenced by field. Thus a complex coupled spectrum may simplify somewhat as is shown in Figure 4.28, for brucine, in spectra obtained at 250 MHz and 600 MHz.

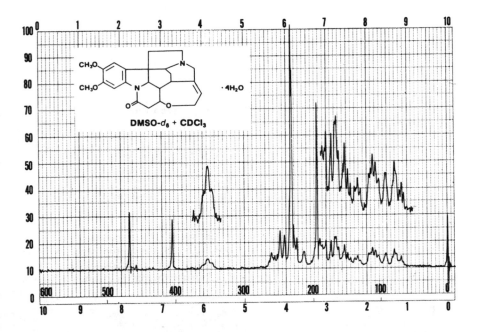

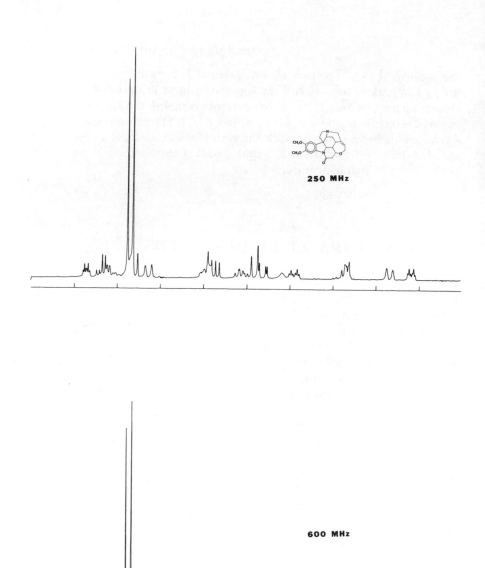

250 MHz

600 MHz

Figure 4.28 Effect of higher fields on proton nmr spectra. Nmr spectra of brucine at 60 MHz (0 to 10 ppm, previous page), 250 MHz (0.4 to 4.8 ppm, top), and 600 MHz (0.4 to 4.8 ppm, bottom). Note the simplification of the multiplets at higher field strengths. Courtesy of Dr. Joseph Dadok and Nikolaus Szeverenyi of Carnegie-Mellon University.

THE USE OF LANTHANIDE SHIFT REAGENTS

A similar effect can be obtained for some compounds using the lanthanide "shift reagents" which form complexes with the polar parts of many molecules. These reagents are actually paramagnetic in nature and thus cause substantial changes in the chemical shifts of groups near the site of the complex, but do not cause the line broadening associated with more common paramagnetic substances.

Two commonly used shift reagents are β-diketone complexes of europium and praeseodymium. Praeseodymium complexes cause upfield shifts of the

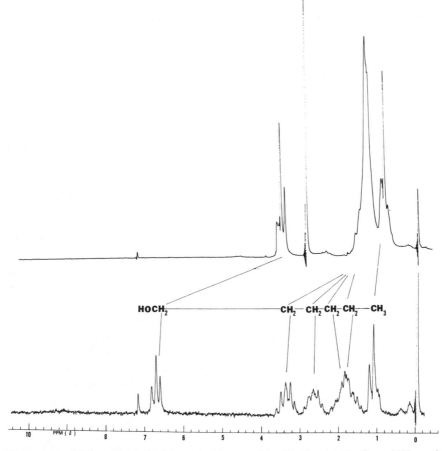

Figure 4.29 Nmr spectra of 1-hexanol: (*a*) normal in $CDCl_3$ and (*b*) after addition of $Eu(fod)_3$.

complexed groups and europium complexes cause downfield shifts. The specific diketones used are usually the dipivalyl-methanato complex (thd) or the heptafluoro-dimethyloctanedionato complex (fod).

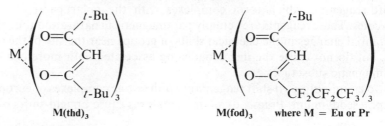

M(thd)₃   M(fod)₃   where M = Eu or Pr

The utility of these shift reagents is illustrated in Figure 4.29 for the spectrum of 1-hexanol in $CDCl_3$. The upper trace is the normal 60 MHz spectrum and the lower trace is the shift reagent spectrum after the addition of 0.2 mg of $Eu(fod)_3$.

SPECTRAL FEATURES OF MAJOR FUNCTIONAL GROUPS

Alkanes

Most alkanes show their resonances between $\delta 0.9$ and $\delta 1.5$. Only very simple alkanes lead to first-order spectra that are readily analyzable. More common are the sort of broad blobs shown in Figure 4.30 for the spectrum of 1-phenyl-decane. The methyl multiplet is separated at $\delta 0.88$ and the CH_2 α to the aromatic ring is found at $\delta 2.56$, but the remaining protons are clustered together in a large broad peak around $\delta 1.28$.

More information can sometimes be obtained from alkanes by running the spectra at higher fields so that the chemical shifts spread out, while the coupling constants remain the same. Spectra at 270 or 360 MHz are becoming quite common with higher field spectrometers now in development by several groups. A somewhat similar effect can be obtained in many cases by using the so-called Lanthanide Shift Reagents, which produce large changes in the chemical shifts of nuclei in molecules to which they complex, due to the magnetic moment of the unpaired electron of the lanthanide ion. While many paramagnetic substances also produce extreme line broadenings, the complexes of a few rare earths can be used to induce substantial shifts without excessive broadening.

Cycloalkanes

The larger cycloalkanes have more or less the same chemical shifts as the open chain alkanes: most of the methylene hydrogens are found in the region

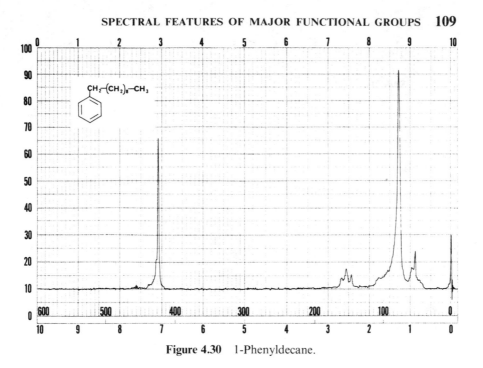

Figure 4.30   1-Phenyldecane.

of $\delta1.2$–$\delta1.4$. Cyclopropanes are an exception, however, having chemical shifts in the range of $\delta0.1$–$\delta0.5$. This great shielding effect is usually ascribed to the anisotropy (uneven electron distribution) of the cyclopropane C—C bonds. This is a rather large effect since, as you may recall, the C—C bonds are considered to be sp^2 according to the Walsh model and would normally cause proton chemical shifts in the $\delta4$–$\delta5$ range. Vicinal couplings (J_{H-C-H}) in cyclopropanes are generally found to be around -4.3 Hz and are more observable than open chain vicinal coupling constants, since substituents in rings often make two vicinal protons nonequivalent. Because of the sp^2 hybridization of cyclopropyl rings, the three-bond geminal coupling constants vary from a maximum of 10 (*cis*) to 4 Hz (*trans*). These factors usually lead to extremely complex upfield spectra indicative of cyclopropyl species. The only substituted cyclopropanes that are singlets, in fact, are those held spiro to another ring or symmetrically substituted.

Cyclohexanes have been studied extensively by nmr spectroscopy since the relationship between bond angle and coupling constant makes nmr an ideal tool for the study of conformation. Thus the diaxial 1, 2 coupling has been found to be 8–10 Hz (180°) and the axial-equatorial and diequatorial coupling (60°) to be 2–3 Hz. Both are in agreement with the Karplus equation.

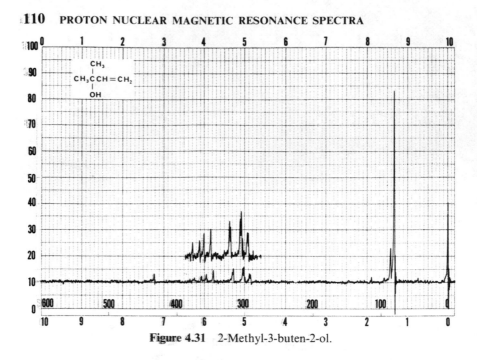

Figure 4.31 2-Methyl-3-buten-2-ol.

If a puckered ring is flipping rapidly, we see only a single line for protons that can interconvert between axial and equatorial, but if the conformation is locked by a large group or another ring, two separate lines will be observed. For intermediate cases, where the interconversion rate is slow on the nmr time scale, we might see anything from broadening of the two lines to broadening of the singlet depending on the rate involved.

Alkenes

Alkenes show resonances in the region of $\delta 4.5-\delta 6.0$ and are usually complex multiplets because of their more efficient long-range coupling abilities. Not only do large H—C=C—H couplings (10–17 Hz) occur commonly, but the allylic coupling (H—C—C=C—H) is usually at least -2 Hz, and homo-allylic couplings (H—C—C—C=C—H) of 1–2 Hz are not uncommon. Thus highly complex splitting patterns in this region are nearly always due to alkenes rather than deshielding by some strongly electronegative group. A typical pattern showing some allylic coupling is shown in Figure 4.31 for 2-methyl-3-buten-2-ol.

Alkynes

Because of the cylindrical anisotropy of the triple bond system, the hydrogens attached to alkynes are actually shielded more than we would expect from

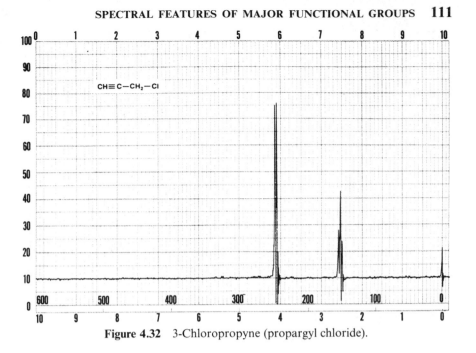

Figure 4.32   3-Chloropropyne (propargyl chloride).

the electron withdrawing power of the triple bond. In Figure 4.32 we see the expected two lines for 3-chloropropyne, but upon examining their intensities we see that the methylene —CH_2 is farther downfield ($\delta 6.13$) than the alkyne CH ($\delta 2.53$). This is confirmed by the splittings of the peaks: the CH_2 is split into a doublet by the CH and the CH into a triplet by the CH_2. Note that this is a four-bond allylic coupling of nearly -3 Hz.

Interior alkynes do not, of course, show these terminal alkyne C—H resonances, but CH's adjacent to the triple bond will be deshielded somewhat and often show coupling with groups across the triple bond. These are analogous to homoallylic couplings and are usually on the order of 1–2 Hz. For example, in the spectrum of the severe poison oxotremorine in Figure 4.33, the downfield multiplet at $\delta 4.18$ is due to the two slightly different CH_2's on either side of the triple bond, coupling with each other. Their chemical shifts are very similar, however, and they overlap into the same peak.

Alcohols

The —OH proton in alcohols can come at virtually any chemical shift, depending on the solvent and any hydrogen bonding that may occur. Because of this hydrogen bonding the —OH protons are nearly always singlets since they exchange so fast that they see only an average magnetic field from other nuclei and induce only an average field on other nuclei. Couplings to

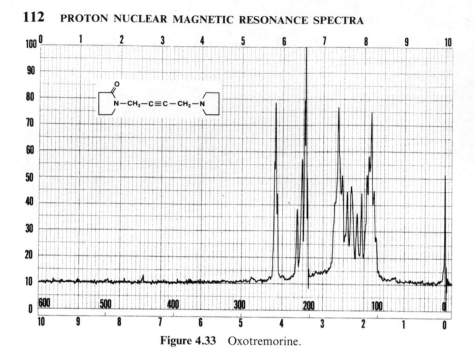

Figure 4.33 Oxotremorine.

alcohol protons can be observed, however, if the alcohol is dissolved in di-methyl sulfoxide (DMSO), where hydrogen bonding to the solvent occurs and little exchange can take place. This is illustrated in Figure 4.34. Alcohol OH peaks are often rather broad and sometimes may be lost completely as very broad low-intensity humps.

The CH_n groups α to the alcohol OH are usually deshielded by the electro-negativity of the oxygen and are found at $\delta 3.4$–$\delta 4.0$. This usually makes their coupling interactions first order with other adjacent protons which are more alkane-like, and the number of hydrogens β to the OH can often be deduced.

Ethers

Protons on carbons next to ether oxygens are usually observed in the range of $\delta 3.3$–$\delta 4.0$. Coupling is seldom observed across ether oxygens as is illus-trated by the spectrum of 1,1-dichloromethyl methyl ether in Figure 4.35.

Alkyl Halides

Protons attached to a carbon bearing one halogen are usually observed in the range of $\delta 2.0$–$\delta 4.0$, but they may be substantially more deshielded if attached

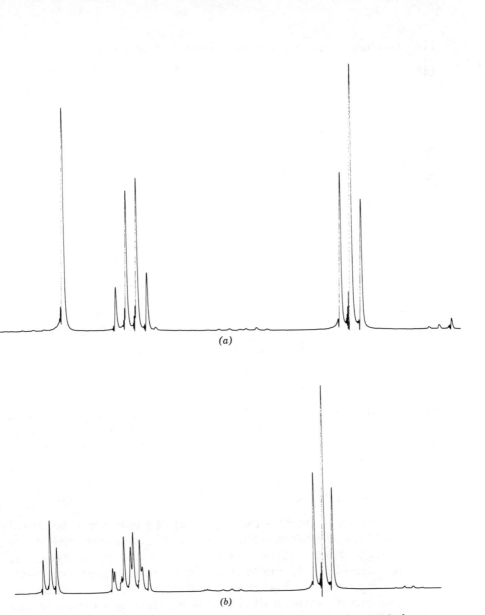

Figure 4.34 Spectra of absolute ethanol in (a) CDCl$_3$ and (b) DMSO-d_6.

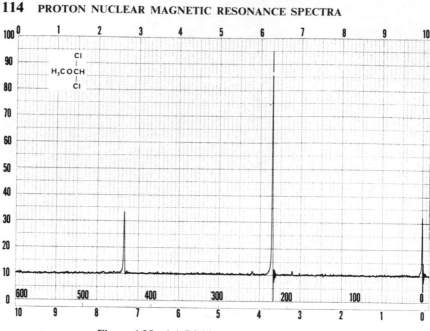

Figure 4.35   1,1-Dichloromethyl methyl ether.

to a carbon bearing two halogens. For example, methylene chloride, CH_2Cl_2, has a chemical shift of $\delta 5.32$, and the hydrogen attached to the carbon bearing the two halogens and the oxygen is deshielded to $\delta 7.33$ in the spectrum of the dichloromethyl methyl ether in Figure 4.35.

Ketones and Aldehydes

Protons on carbons adjacent to carbonyl groups are generally deshielded to the $\delta 2.0-\delta 2.7$ range. While little four-bond coupling across carbonyls is ever observed, the coupling across carbonyls to aldehyde protons is fairly efficient and in the range of 2–3 Hz. Furthermore, the aldehyde protons are strongly deshielded and occur in the range of $\delta 9-\delta 10$. This latter effect is illustrated in the spectrum of acetaldehyde (Figure 4.36), showing a narrow upfield doublet for the methyl group at $\delta 2.21$ and a downfield quartet for the CHO proton at $\delta 9.83$.

Carboxylic Acids

Protons on carbons α to the carbonyl in carboxylic acids have roughly the same chemical shifts as those on aldehydes and ketones ($\delta 2.0-\delta 2.7$). The highly acidic COOH proton has very little electron density nearby and is one

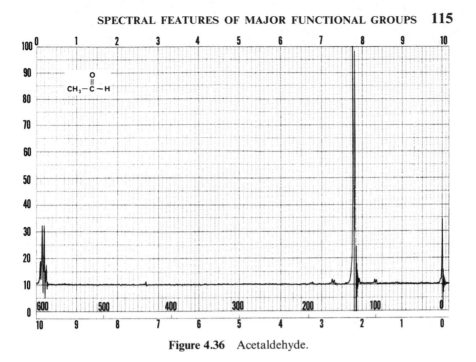

Figure 4.36 Acetaldehyde.

of the more deshielded protons in pmr spectroscopy, coming in the range of $\delta10$–$\delta12$. Since it undergoes substantial hydrogen bonding, coupling to this proton is never observed. This is illustrated in the spectrum of the vile-smelling butyric acid, a principle component in vomit, shown in Figure 4.37. The 550 Hz offset needed to bring this COOH proton on scale means that its actual shift is [60 MHz (1.82 ppm) + 550 Hz]/60 MHz, or $\delta10.99$.

Amines

Protons on carbons α to NH_2 and NR_2 groups are only deshielded to the region $\delta2.0$–$\delta2.8$. The NH protons exchange rapidly because of hydrogen bonding and are always observed as singlets. They can be found anywhere in the range of $\delta1.0$–$\delta5.0$, depending on the solvent. This is illustrated in Figure 4.38 for the spectrum of propylamine, where the NH_2 protons are observed overlapping those of the CH_3 group, at $\delta1.04$.

Aromatic Compounds

Aromatic protons are usually found in the range $\delta6.0$–$\delta8.5$. If the aromatic ring is substituted with groups of only slightly different electronegativity

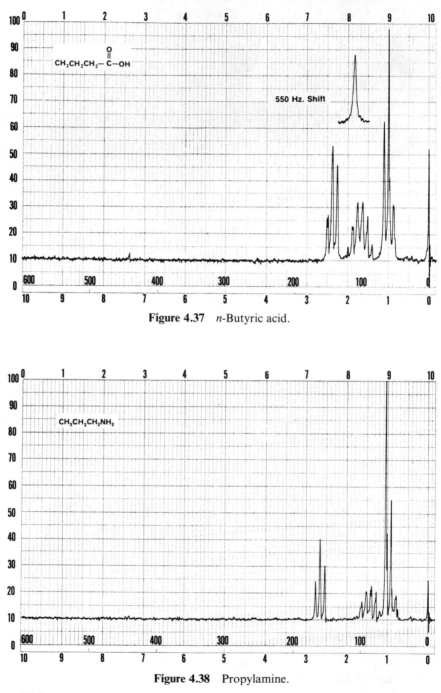

Figure 4.37 *n*-Butyric acid.

Figure 4.38 Propylamine.

than H, the remaining protons usually are found in a single broad peak. This was illustrated in Figure 4.5 for toluene and xylene.

Benzylic protons are usually found in the range of $\delta 2.3$–$\delta 3.0$ and occasionally show couplings to the aromatic protons of 0.3–0.6 Hz. Couplings of 0.5 Hz to the protons *para* to the substituent are sometimes observed and are apparently due to the rather efficient transmission of spin information through the aromatic π-system.

Disubstituted rings generally provide more complex spectra in the aromatic region. Those which are *ortho* and *meta* substituted provide no easily recognizable features other than the fact that they are not monosubstituted. This is illustrated in Figure 4.39 for 2-fluorobenzyl chloride.

Para-disubstituted benzene rings often produce a pattern of two doublets leaning toward each other, characteristic of the $AA'BB'$ (almost A_2B_2) system. This pattern is most recognizable when there are substantial electronegativity differences between the two groups. When they are identical, the pattern becomes a singlet, since the four hydrogens are then virtually equivalent. A typical pattern is observed in the spectrum of 1-chloro-4-iodobenzene shown in Figure 4.40.

Organometallics

Organometallic compounds such as organolithium and Grignard reagents have substantial carbanion character and are thus greatly shielded by this negative charge. The CH_2MgX protons of Grignard reagents are often observed around $\delta 0.5$ and those of organolithium compounds as high as $\delta - 1.3$. Organomercury and organocadmium compounds are substantially more covalent and are all found below $\delta + 1.0$.

The carbanion character of lithium compounds and Grignard reagents has been extensively studied, and it has been found that they rather resemble amines in having an unshared pair of electrons that can invert like an umbrella. The rate of this inversion has been measured for a number of species by nmr spectroscopy.[13]

COMPUTER CALCULATION OF NMR SPECTRA

Unlike most other chemical measurement techniques, the theory of complex nmr spectra is fully developed, and the characteristics of the functions that operate even in complex cases are fully known. Thus a number of programs have been written for the calculation of nmr spectra from the chemical shifts and coupling constants, which can then be displayed and plotted out for

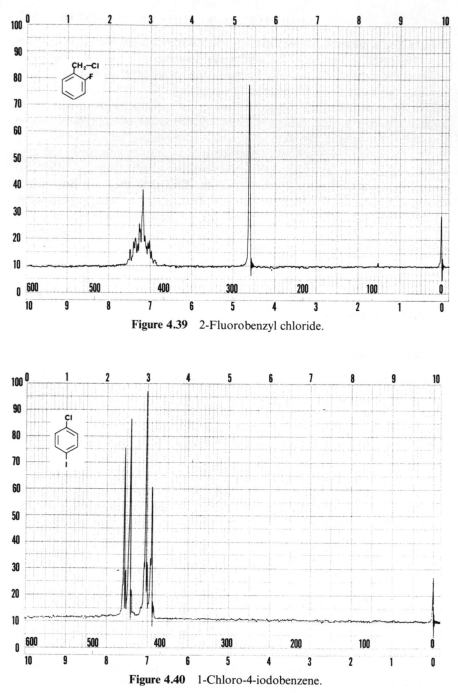

Figure 4.39 2-Fluorobenzyl chloride.

Figure 4.40 1-Chloro-4-iodobenzene.

118

comparison with experimental spectra. Computer programs have been written for the PDP–8/I, PDP–8/e, Nicolet 1080, 1180, PDP–11, and Aspect–2000, which are fundamentally similar in their operation and are described below. Programs are also available for the Varian 620/i and V77 and for the Texas Instruments 990 series.

To use these programs, load and start them as necessary from disk or paper tape. The programs will type out their name and ask for the number of nuclei in the system. They usually allow between two and six or seven spin-$\frac{1}{2}$ nuclei.

NMRSPEC
NUMBER OF SPINS =

Answer with the number between 2 and 7 followed by a Return. The programs then ask for the spectral width to be displayed, any offset from 0, and the spectrometer frequency in MHz.

SWEEP WIDTH =
OFFSET =
SPECTROMETER FREQUENCY (MHz) =

Answer each question appropriately and go on to enter the chemical shifts

$V(1) =$
$V(2) =$
etc.

and the coupling constants

$J(1, 2) =$
$J(1, 3) =$

etc.

When the parameters have all been entered the program will begin the calculation. It will be virtually instantaneous for a two- or three-spin case, but will require a number of seconds for the four- and five-spin cases and a number of minutes for the six- and seven-spin cases since larger and larger matrices must be diagonalized as part of the calculation. When the calculation is done, a display of the stick figure representation of the spectrum will appear on the scope. This can be replaced by a conventional spectrum including line width by typing L and entering a number for the line width in Hz, usually around 0.3–1.0. The resulting spectrum can be expanded by changing the width and offset and plotted as either a stick or Lorentzian line shape spectrum. Furthermore most programs also allow individual parameters to be changed and the spectrum recalculated to produce one more like the experimental one.

ITERATIVE FITTING OF NMR SPECTRA

Rather than blind trial and error fitting of experimental complex spectra to chemical shifts and coupling constants, it would be preferable to have a computer program that would calculate a set of shifts and couplings which produced a *best fit* of experimental data to theoretical parameters. There are two such programs written in FORTRAN for large scale computers, named NMRIT and LAOCOON III.[14,15] The former program iterates the energy levels for a best fit with the observed transitions and the latter program iterates directly on the transitions. An assembly language version of LAO-COON III has been prepared for the Nicolet 1080 and 1180 series, named ITRCAL, and a similar version is available for the Aspect–2000.

The LAOCOON III program (which stands for Least-Squares Adjustment on Calculated of Observed Nmr spectra, revision III) is designed for card input of guessed shifts and couplings; when the calculated spectrum is reasonably close to the observed one, assignment of the calculated lines to the observed lines is made and the program then iterates for a best fit between the two. The original version of this program required that card input of the guessed spectra be prepared repeatedly until a spectrum close to the experimental one was produced and then a set of transitions were punched on cards. These cards are collected and assigned frequencies punched into them. The cards are then reread with appropriate control cards signifying iteration to make the assigned frequencies best fit the calculated frequencies.

With the advent of timesharing systems, an interactive conversational version of this program for use on the DECsystem–10 has been prepared and a listing is given in Appendix I. The card-input, printer–output system has been replaced by dialog at the timesharing terminal with an optional display if a graphics terminal is provided. To use this version of the program, log into the system and type

RUN LAOCN3

The program will load and start, typing out

LAOCN3

NUMBER OF SPINS=

Answer with any number between 2 and 7 followed by a Return. A number outside this range will cause the program to halt and exit to the monitor. The program will next print

ENTER TITLE:

Enter a title of up to 60 characters, which is to be printed out in the final listing followed by a Return.

MIN. FREQ. =

MAX. FREQ. =

Enter the maximum and minimum frequencies of interest in the calculation. LAOCN3 has the ability to allow you to calculate couplings between different kinds of nuclei such as ^{1}H and ^{31}P, without making their chemical shifts differ on the order of megahertz, by having you declare the names of the nuclei in the same order as the chemical shifts. For example, if you were going to calculate the splittings in the six-spin system CH_3OPH_2, you would enter the names of the nuclei in the order HHHPHH, and then enter the chemical shifts in the same order. For most cases, the nuclei will all be of the same kind, and the question

ENTER LETTERS FOR EACH NUCLEUS:

can be answered with a series of H's: one for each spin in the system. The program then asks for the chemical shifts (V's) and coupling constants (J's):

$V(1) =$
$V(2) =$ etc.

$J(1, 2) =$
$J(1, 3) =$

etc.

Enter the value for each shift and coupling constant as the question is printed out followed by a Return.

When all the shifts and coupling constants have been entered, the program will want to know if you are iterating. Answer the question N if a direct calculation is being performed and Y if iteration is being performed.

ITERATING? (Y or N):

If you answer N the program will list the calculated transitions and intensities on the line printer and will create a disk file of them if you desire:

CREATE FILE OF TRANSITIONS? (Y or N):

If you answer N the program will restart and ask for new data. If you answer Y it will ask for the six-character file name under which the data is to be stored:

DISK FILE NAME=

All such files will have the ".DAT" extension unless one is specifically entered. The program will create the file and then restart:

LAOCN3
NUMBER OF SPINS=

To stop the program enter a zero number of spins. It will stop and ask if the output should be printed on the line printer. If you answer Y, a listing will be produced, if N, the program will exit to the monitor.

Iteration Using LAOCN3

In order to iterate your experimental data for a best fit with the theoretical calculation, you must assign your experimental frequencies to the closest calculated ones. Once you have such an assignment, you can either add this data to the transition list in the file you created or actually enter them at the terminal during the iteration process.

Let us suppose that we have a spectrum of malic acid such as is shown in Figure 4.41. The formula for malic acid is

$$HOOC-CH_2-CHOH-COOH$$

and the CH_2 and the CHO protons form an ABX system from about 1–4 ppm. The frequencies shown are measured relative to the right-most line. Since malic acid has a chiral center at the CHOH carbon, the adjacent CH_2 protons are nonequivalent and thus form the AB part of the ABX system. ABX systems can be analyzed by hand with some difficulty, but quite quickly, using computer programs as we will demonstrate below.

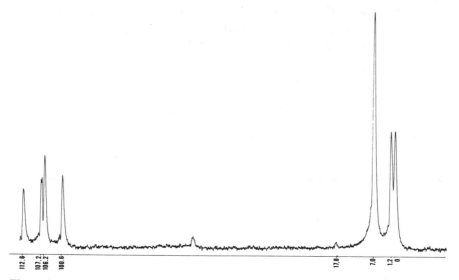

Figure 4.41 Malic acid ($HOOC-CH_2-CHOH-COOH$) in D_2O. The peak measurements are given in Hz from the upfield peak.

In order to use the iterative capabilities of LAOCOON III, we must first guess a set of chemical shifts and coupling constants that will produce a set of transitions and intensities. We then assign these to experimental lines and iterate for a best fit.

We will then start by guessing a set of shifts and coupling constants based on the spectrum and our knowledge of the structure of the compound. Looking at the upfield AB part of the spectrum, we might guess chemical shifts of 1 and 7 Hz and about 107 Hz for the X proton. The coupling constants J_{13} and J_{23} are simple alkyl couplings and we might estimate them arbitrarily at about 7 Hz. There will also be a coupling between the two AB protons which will be important since they are nonequivalent. Without looking back in our text we might remember that such couplings have a magnitude of around 12 Hz and forget the correct sign. Let us proceed with $J_{12} = 12$ Hz and see what results we get.

We might well start by using a minicomputer program to see if these are reasonable values, since such programs usually display the data immediately. If none is available, we might simply run the LAOCN3 program in non-iterative mode to get a list of theoretical transitions and intensities. Once this list has been created as a disk file we can list it on the terminal and examine the frequencies of the lines relative to the observed experimental frequencies.

Let us start the program and enter the guessed shifts and coupling constants:

```
LAOCOON3
NUMBER OF SPINS= 3
ENTER TITLE: MALIC ACID
MIN. FREQ.= -10
MAX FREQ.= 120
ENTER LETTERS FOR EACH NUCLEUS: HHH
V(1)=1
V(2)=7
V(3)=107
J(1,2)= 12
J(1,3)= 7
J(2,3)= 7
```

Since we are iterating, we answer Y and then indicate that we will enter the assigned transitions at the terminal rather than from a disk file. If two transitions are to be assigned to the same experimental frequency, we can use a

```
ITERATING? (Y OR N): Y
READ IN ASSIGNED TRANS. FROM DISK?(Y OR N): N
MIN. INTENSITY= 0
    7      -5.440     0.108
   11      -0.204     1.770    0
   14       0.974     1.755    1.2
    1       6.576     2.019    7
    4       8.179     2.034    7
   12      12.990     0.101
    2      20.195     0.110   17.8
    8      93.831     0.000
   13     100.246     1.143  100.6
    9     107.025     1.001  106.2
    5     107.450     0.985  107.2
    3     114.230     0.871  112.6
```

ditto mark (") the second and succeeding times to indicate that the last entered value is to be repeated. Note that if we are making no assignment for a given transition, we simply type a Return. It is not necessary to assign all of the transitions.

We then specify which parameters are to be varied and whether they are to be varied in sets. The program allows as many as six parameters (shifts, couplings) to be varied *as a group*, by entering them on the same line. To specify that shifts are to be varied, we enter a single digit number for each shift (1–7) in the order they were originally entered to the program. To specify that couplings are to be varied, we enter two digit numbers, so that J_{34} is 34. If we want to specify that shifts $V(2)$ and $V(3)$ should always be the same, we simply enter them on the same line as one parameter set: 2, 3. If we were to specify that J_{23} and J_{34} were to be varied together, we enter the two digits on the same line: 23, 34. If no parameters are to be varied together as is true in this case, we enter each of them on a separate line. Parameters will only be varied if they are entered in this list. To terminate the entry of new sets, we enter a zero for the last set.

```
ENTER PARAMETERS TO BE VARIED:
  1    1
  2    2
  3    3
  4    12
  5    13
  6    23
  7
```

When the calculation is finished, the program prints out the final values on the terminal, and if a graphics terminal is being used, it draws the final spectrum. If a normal terminal is being used, you can always return to the minicomputer noniterative program to plot out the results. The results of this fit are shown in Figure 4.42. It is clearly not too good, both by the size of the errors and the lack of visual agreement between the experimental and calculated data.

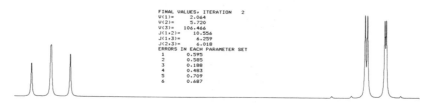

```
FINAL VALUES, ITERATION   2
V(1)=      2.064
V(2)=      5.720
V(3)=    106.466
J(1,2)=     10.556
J(1,3)=      6.259
J(2,3)=      6.018
ERRORS IN EACH PARAMETER SET
  1        0.595
  2        0.585
  3        0.188
  4        0.483
  5        0.709
  6        0.687
```

Figure 4.42 Iterated fit of malic acid by LAOCN3 using a positive geminal coupling constant. The peak positions and intensities do not closely resemble the actual spectrum.

```
LAOCOON3
NUMBER OF SPINS= 3
ENTER TITLE: MALIC ACID WITH -J
MIN. FREQ.= -10
MAX FREQ.= 120
ENTER LETTERS FOR EACH NUCLEUS: HHH
V(1)=1
V(2)=7
V(3)=107
J(1,2)= -12
J(1,3)= 7
J(2,3)= 7
ITERATING? (Y OR N): Y
READ IN ASSIGNED TRANS. FROM DISK?(Y OR N): N
MIN. INTENSITY= 0
    1        -5.219      0.115
   15        -0.439      1.764     0
   12         1.210      1.775     1.2
    7         6.765      2.025     7
    2         7.989      2.014     7
   14        13.194      0.093
    4        19.973      0.118    17.8
    8        93.817      0.000
   13       100.246      1.143   100.6
    5       107.026      0.997   106.2
    9       107.450      0.989   107.2
    3       114.230      0.871   112.6
ENTER PARAMETERS TO BE VARIED:
 1   1
 2   2
 3   3
 4  12
 5  13
 6  23
 7
```

Figure 4.43 Program entries for iteration of malic acid using a negative geminal coupling constant.

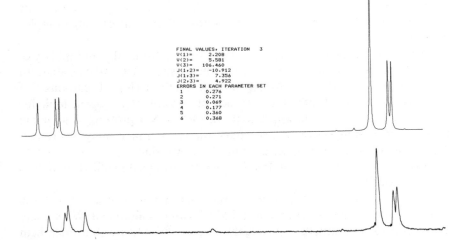

Figure 4.44 Iterated and experimental spectra of malic acid using a negative geminal coupling constant.

125

In such cases, it is always a good idea to vary the sign of any coupling constants that might conceivably be in doubt. Here we will vary the sign of $J(1, 2)$ and make a new iterative fit. The program entries are shown in Figure 4.43 and the resulting fit in Figure 4.44. This is an excellent fit. It should be noted that changing the signs of coupling constants does not vary the frequencies of the assigned lines, only their intensities, and thus there will be two fits depending on the sign of the coupling constants.

NMR SPECTRAL PROBLEMS

The problems that follow vary in complexity from relatively simple to somewhat challenging. Most of them are nearly first-order spectra since it is assumed that complex analyses require computer assistance and accurate peak positions and intensity data which may not be available in routine nmr spectra presented to you to interpret.

To attack these problems, start with the molecular formula and determine the degree of unsaturation and/or number of rings. Molecular formulas rather than empirical ones are given since, as we will see later, they are readily obtainable from mass spectral data.

Decide between unsaturation and rings by examining the chemical shifts. Assume that there are aromatic rings if there are lines between $\delta 7$ and $\delta 8.5$. Assume that there may be double bonds if there are lines between $\delta 4.5$ and $\delta 6.0$, unless a carbon could have two or more electronegative groups such as halogens or oxygens attached. A complex pattern in the vinyl region usually assures you that one or more double bonds are present causing complex couplings.

Determine the major functional groups by examining the multiplicity of the peaks. Triplets usually mean CH_2's are nearby, quartets indicate adjacent methyl groups, and doublets usually indicate methine (R_3CH) carbons.

If the formula indicates the presence of oxygen or nitrogen, look for slightly broadened singlets which could be OH or NH protons. If none are found, then carbonyls, ethers, or tertiary amines or amides may be present. They usually will cause spectral simplifications since they will separate groups from each other and shift adjacent CH_n groups downfield enough to assure first-order spectra.

With these data draw the various structures that are consistent with the formula and see which ones match and which can be eliminated. When there are several possible structures having the correct multiplicities, consider them in terms of the likely chemical shifts of various arrangements of groups to make your final assignments.

PROBLEMS

Problem 4.1 $C_{14}H_{14}$

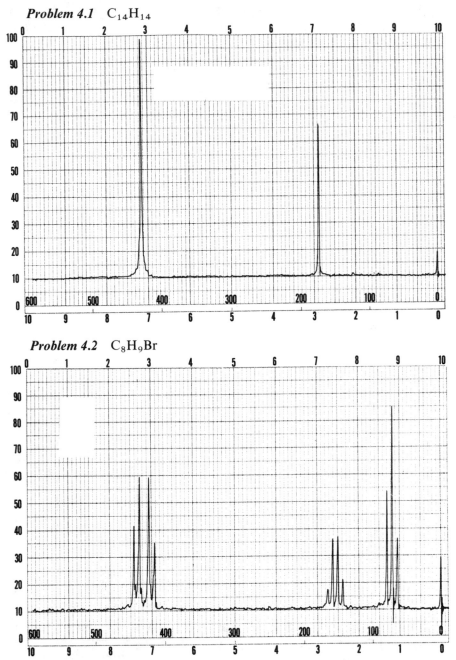

Problem 4.2 C_8H_9Br

Problem 4.3 $C_4H_6Cl_4$

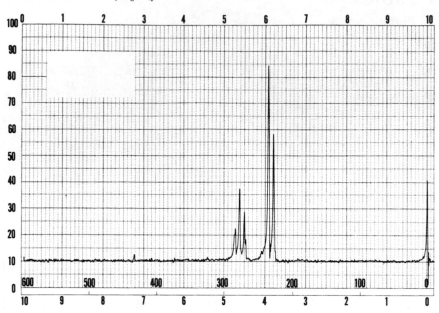

Problem 4.4 $C_7H_{14}O$

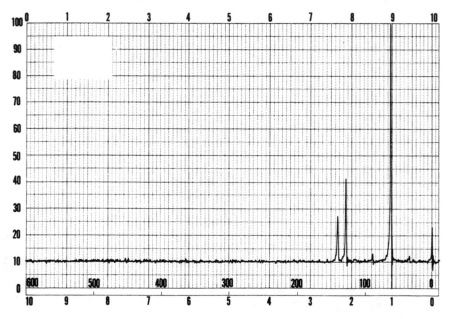

Problem 4.5 $C_4H_6Cl_2$

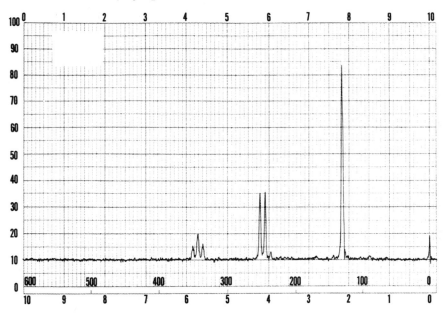

Problem 4.6 C_4H_8O

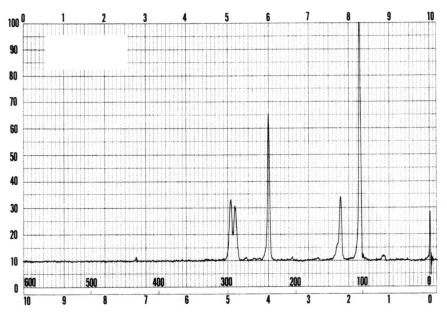

Problem 4.7 C_8H_8

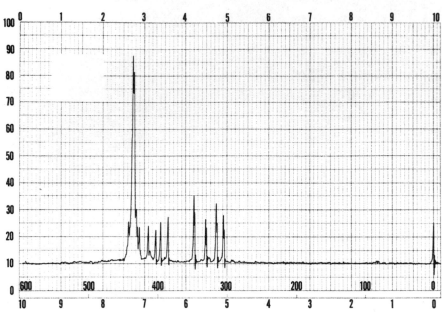

Problem 4.8 C_5H_9ClO

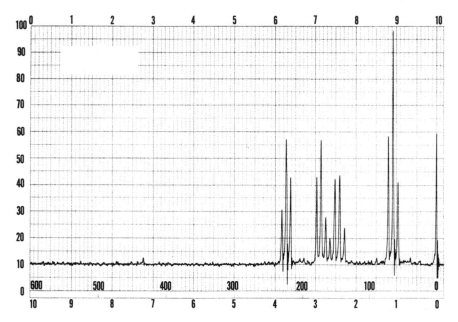

Problem 4.9 $C_3H_5ClO_2$

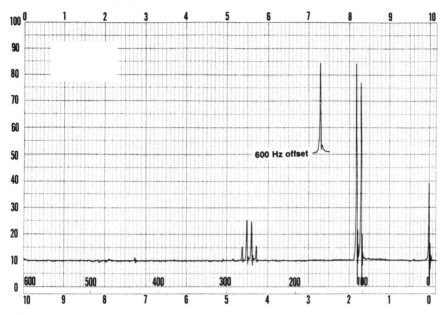

600 Hz offset

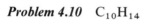

Problem 4.10 $C_{10}H_{14}$

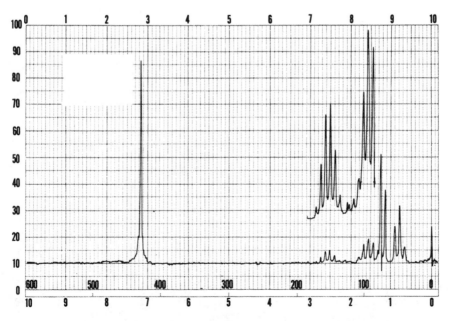

Problem 4.11 C_8H_8O

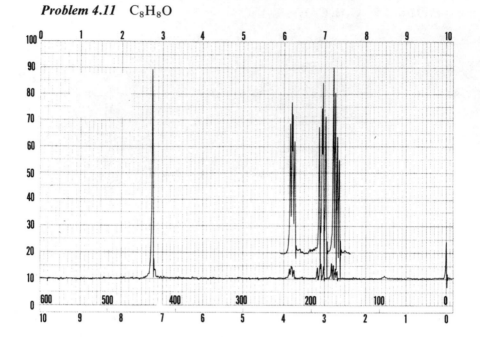

Problem 4.12 $C_{11}H_{16}O$

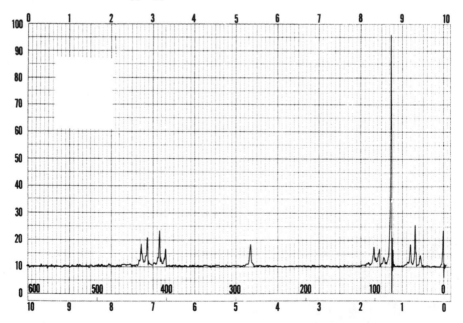

Problem 4.13 $C_5H_8F_4O$

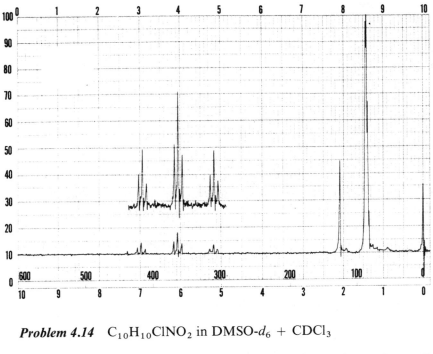

Problem 4.14 $C_{10}H_{10}ClNO_2$ in DMSO-d_6 + $CDCl_3$

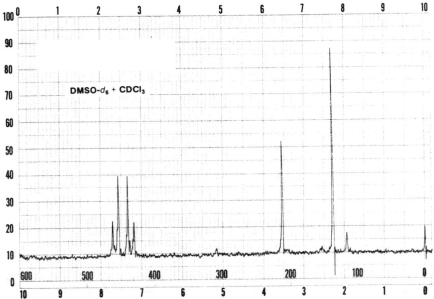

DMSO-d_6 + CDCl$_3$

RECOMMENDED READING

Ault, A., and G. O. Dudek, *NMR—An Introduction to Proton Nuclear Magnetic Resonance Spectroscopy*, Holden-Day, San Francisco, 1976.

Becker, E. D., *High Resolution Nmr*, Academic Press, New York, 1969.

Bovey, F. A., *Nuclear Magnetic Resonance Spectroscopy*, Academic Press, New York, 1969.

Dyer, J. R., *Applications of Absorption Spectroscopy of Organic Compounds*, Prentice-Hall, New York, 1975.

Jackman, L. M., and F. A. Cotton, Eds., *Dynamic Nuclear Magnetic Resonance Spectroscopy*, Academic Press, New York, 1975.

Jackman, L. M., and S. Sternhell, *Applications of Nuclear Magnetic Resonance Spectroscopy in Organic Chemistry*, Pergamon, Oxford, 1969.

Lambert, J. B., *et al.*, *Organic Structural Analysis*, MacMillan, New York, 1976.

Pavia, D. L., G. M. Lampman, and G. S. Kriz, *Introduction to Spectroscopy*, W. B. Saunders, Philadelphia, 1979.

Pople, J. A., W. G. Schneider, and H. J. Bernstein, *High-resolution Nuclear Magnetic Resonance*, McGraw-Hill, New York, 1959.

SPECTRAL COLLECTIONS

Bhacca, N. S., D. P. Hollis, L. F. Johnson, and E. A. Pier, "Nmr Spectral Catalog," Varian Associates, Palo Alto, 1963.

Pouchert, C. J., and J. R. Campbell, "The Aldrich Library of Nmr Spectra," Vol. 1–11, Aldrich Chemical Co., Milwaukee, 1974.

REFERENCES

1. H. Spiesecke and W. G. Schneider, *J. Chem. Phys.*, **35**, 722 (1961).
2. L. M. Jackman et. al., *J. Am. Chem. Soc.*, **84**, 4307 (1962).
3. E. D. Becker, R. B. Bradley, and C. J. Watson, *J. Am. Chem. Soc.*, **83**, 3743 (1961).
4. G. Fraenkel et. al., *J. Am. Chem. Soc.*, **82**, 5846 (1960).
5. T. J. Katz, M. Rosenberger, and R. K. O'Hara, *J. Am. Chem. Soc.*, **86**, 249 (1964).
6. A. D. Buckingham and K. A. McLachlan, *Proc. Chem. Soc.*, **144** (1963); L. C. Snyder, *Bull. Am. Phys. Soc.*, Ser. II, **10**, 358 (1965).
7. M. Karplus, *J. Chem. Phys.*, **30**, 11 (1959).
8. J. A. Pople and A. A. Bothner-By, *J. Chem. Phys.*, **42**, 1339 (1965).
9. M. Barfield and B. Chakrabarti, *Chem. Rev.*, **69**, 757 (1969); F. A. L. Anet et. al., *J. Am. Chem. Soc.*, **87**, 5249 (1965).
10. S. Sternhell, *Rev. Pure Appl. Chem.*, **14**, 15 (1964), as cited in J. B. Lambert et. al., reading 7.

11. E. D. Becker, *High Resolution Nmr*, Academic Press, New York, 1969, pp. 149 ff.
12. M. Saunders, P. Schleyer, and G. A. Olah, *J. Am. Chem. Soc.*, **86**, 5680–5681 (1964).
13. G. Fraenkel, C. Cottrell, and D. T. Dix, *J. Am. Chem. Soc.*, **93**, 1704 (1971).
14. S. Castellano and A. A. Bothner-By, *J. Chem. Phys.*, **41**, 3863 (1963); S. Castellano in *Computer Programs for Chemistry*, Vol. 1, D. F. DeTar, Ed., Benjamin, New York, 1968.
15. P. Diehl, H. Kellerhals, and E. Lustig, *Computer Assistance in the Analysis of High-Resolution Nmr Spectra*, Vol. 6 of " Nmr—Basic Principles and Progress," Springer-Verlag, New York, 1972.

CHAPTER FIVE
FOURIER TRANSFORM
NMR SPECTROSCOPY

Thus far we have considered nmr as a single scan experiment where we sweep slowly through the spectrum, varying either the field or the frequency, and plot out the absorptions. In fact, it is perfectly possible to obtain spectra of very much less material by several techniques using a small minicomputer attached to our spectrometer. We will discuss these techniques in this chapter after a brief introduction to the computer itself to give us a vocabulary with which to talk about the experiments.

SIGNAL AVERAGING

Let us suppose that we have a fairly dilute sample containing weak resonances that cannot be observed by plotting the spectrum out, even at conditions of high radio frequency field intensity and high amplifier gain. This noisy spectrum consists of a weak but constant signal and a large amount of noise which is random in nature. If, instead of plotting this spectrum out, we read it into the memory of a small digital computer, we can perform *signal averaging* and enhance our signal-to-noise (S/N) ratio. This process is accomplished by instructing the computer to sample the spectrum at regular time intervals, convert these voltages into numbers, and add them into memory.[1]

On the first scan, the memory is set to zero for the entire array to contain the spectrum, and data points are simply *placed* into memory as the spectrometer scans through the spectrum. Subsequent scans are started at exactly the same frequency and data are sampled at exactly the same time intervals so that the same frequency information goes into the same memory locations each time. As this summation continues, the signal that is weak but coherent grows with each successive scan

$$S = k_1 B \tag{5.1}$$

where S represents the signal and B is the number of scans.

136

There is a great amount of noise in each scan as well, however, but the noise is random and will not grow at the same rate as the signal. In fact, if the noise has a gaussian distribution around zero volts, the noise will cancel out some of the time and will grow at a rate proportional to the *square root* of the number of scans

$$N = k_2 B^{1/2}. \tag{5.2}$$

Dividing Eqs. 5.1 by 5.2, we find that

$$S/N = KB^{1/2} \tag{5.3}$$

where $K = k_1/k_2$ and we can conclude that the S/N will grow at a rate proportional to the square root of the number of scans. This will hold as long as the numbers being added into the computer's memory do not become too large for the number of bits allotted to them, called the computer's *word length*.

Thus we see that we can enhance the S/N by adding successive scans until the peaks of interest " grow " out of the noise. Now, if we have an initial S/N of s, the final S/N will be given by

$$S = sB^{1/2} \tag{5.4}$$

and thus it will take 100 scans, for example, to gain a 10-fold improvement in S/N. This is illustrated in Figure 5.1. In ordinary swept or continuous wave (cw) nmr, this can be something of a problem, because we must scan fairly slowly to keep the spectral lines from ringing into each other and to maintain high resolution. These so-called " slow passage conditions "[2] are satisfied by sweeping at about 1 Hz/sec or slower. Thus it will take about 500 sec to examine a 500 Hz spectral width. It does not take much calculation to realize that to resolve an exceedingly weak spectrum may require hours or even days of signal averaging. This is seldom practical both from the standpoint of instrument demand and instrument stability.

For the solution of this problem, let us perform the thought experiment of attaching two radio frequency sources to the spectrometer and sweeping through half of the spectrum with each one simultaneously. We can then take the outputs of two sets of receiver coils and connect them to two inputs to our computer and scan the spectrum twice as fast. This will increase the rate at which we can signal average by a factor of 2.

We need not stop with just two sources and two detectors. We could, at least as a thought experiment, sweep with, say, 1000 oscillators and connect each to just one point in the computer's memory. Then we could obtain our spectrum instantaneously. The tremendous increase in S/N per unit time is accompanied by essentially an infinite increase in cost, and the idea of building such spectrometers has been abandoned.

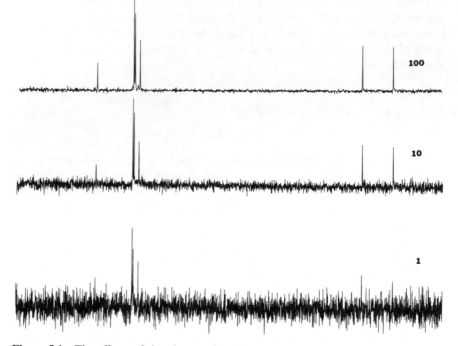

Figure 5.1 The effects of signal averaging. The proton nmr spectrum of dilute ethylbenzene after 1 scan, 10 scans, and 100 scans (5° pulses used).

PULSED-FOURIER TRANSFORM NMR SPECTROSCOPY

The idea of the multichannel spectrometer such as we have been describing has not disappeared, however, because the pulsed-Fourier transform spectrometer[3] accomplishes exactly this technique in a much more straightforward fashion. The pulsed-Fourier transform nmr spectrometer is not substantially different from the swept spectrometer. It contains a magnet, which may be permanent, an electromagnet, or a superconducting solenoid, sources of several radio frequencies, a detector, and a minicomputer for carrying out both signal averaging and the Fourier transform. The one primary difference is that instead of the instrument slowly sweeping through the spectrum, it applies a short burst of high powered radiofrequency energy and then triggers the computer to begin sampling the nuclear response from the detector at a relatively high rate of speed. To understand how this

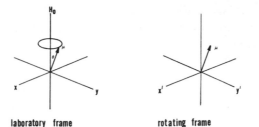

laboratory frame   rotating frame

Figure 5.2 Precession in the laboratory frame and no apparent precession in the rotating frame.

spectrometer works, let us return to our description of the nmr experiment in the rotating frame as shown in Figure 5.2.

Recall that the rotating coordinate system has been used primarily to simplify the drawing of the precessing of the nuclear magnetization around the applied field H_0. Let us now apply a very high powered rotating magnetic field in the x–y plane. In the rotating coordinate system this is easy to draw, as we see in Figure 5.3, where we have labeled this additional field as H_1. This is exactly the same sort of H_1 that we talked about for the cw (continuous wave or swept) experiment in Chapter 3 and is really just an rf field, but it is of much greater intensity. The intensity is so great, in fact, that the total magnetization M made up of all nuclear vectors μ is induced to precess around this H_1 in the rotating frame. This is indicated in Figure 5.3 by the dotted line. This precession will last only as long as we leave the high powered rf field turned on. If we leave it on only long enough for the net magnetization to rotate to the x–y plane, we can observe it when we turn off H_1 with our conventional receiver coil.

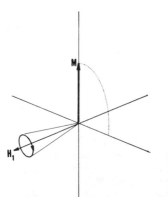

Figure 5.3 Rotation of the net magnetization M about the applied radio frequency field H_1 in the rotating frame.

This magnetization in the x–y plane will be rotating at the Larmor frequency as before, and when H_1 is turned off it can induce a sinusoidal current in the receiver coil placed along the y-axis. This is the nmr signal that we record in this experiment. More important, however, is the fact that this large H_1 can cause the tipping of the magnetization for magnetic moments rotating at frequencies other than the frequency of H_1. In fact, we can excite the entire spectrum at once and observe all of the lines at the same time as a series of sinewaves of different frequencies.

The sample represented in Figure 5.4 has two absorption frequencies represented by vectors v_1 and v_2. The first vector v_1 is rotating at the same frequency as the applied rf field H_1 and is thus stationary in the rotating frame. A second vector v_2 is rotating at a slightly faster frequency and thus " sees " an effective field H_0' relative to H_1. This additional field H_0' is small compared to H_1 and the resultant between H_0' and H_1 is still essentially H_1. This somewhat higher frequency v_2 will also begin rotating around H_1 while it is turned on. In fact, it can be shown that the high powered field H_1 will tip the magnetization of all vectors having frequencies such that

$$\gamma H_1 \gg 2\pi(v - v_0) \tag{5.5}$$

where γ is the magnetogyric ratio for that nucleus, v_0 is the frequency of the pulse, v is the frequency that is excited, and H_1 is the pulse power.

This then is the analog of our multichannel nmr experiment. By applying a brief high powered H_1 in the form of an rf pulse, we can excite a whole range of nmr resonance frequencies. In fact, for most nuclei we can excite the entire chemical shift range of interest at once. Typically the pulse parameters for this experiment are an rf pulse lasting 1–100 μsec and having a power of 10–100 W, depending on the type of spectrometer and probe configuration.

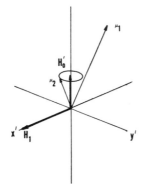

Figure 5.4   Effect of the applied radio frequency field H_1 on nuclei precessing at the same frequency as the rotating frame (v_1) and at a slightly different frequency (v_2). In the rotating frame the net H_0 is much less for both frequencies, and even v_2 can be affected by H_1.

If we want to tip the magnetization M into the x–y plane, we want to tip it through 90° and refer to a pulse that can do this as a *90° pulse*. For such a pulse

$$\gamma H_1 t_p = \frac{\pi}{2} \tag{5.6}$$

where t_p is the time of the pulse and H_1 is the pulse power. Combining Eqs. 5.5 and 5.6 we find that

$$t_p \ll \frac{1}{4(v - v_0)}. \tag{5.7}$$

By plugging in a 5000 Hz spectral width we find that the pulse width t_p must be much less than 50 μsec.

In fact, 90° pulses for many iron magnet spectrometers are 15 μsec or less for ^{13}C spectra, which cover about 5000 Hz and only 1 or 2 μsec for proton spectroscopy.

RELAXATION EFFECTS

Once the magnetization of the various nuclei has been tipped into the x–y plane, we can observe it by the current it induces in the receiver coil. However, since this magnetization represents an excited state, it clearly will not stay there for long once we have turned off our perturbing pulse. There are two relaxation effects that contribute to the return of the magnetization to its equilibrium position along H_0. We have already identified them in Chapter 3 as spin–lattice relaxation and spin–spin relaxation, represented by times T_1 and T_2, respectively.

Spin–lattice relaxation is the tendency of the magnetization to return to the z-axis (Figure 5.5), shown in the rotating frame. In the laboratory frame

Spin–lattice relaxation

Figure 5.5 Spin–lattice relaxation in the rotating frame. The magnetization tends to return to alignment with H_0 when the exciting pulse is turned off.

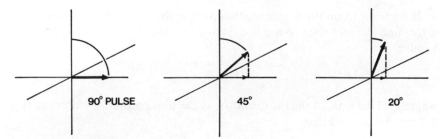

Figure 5.6 Tipping of the magnetization using various length pulses, resulting in 90°, 45°, and 20° flip angles.

it is rather like an umbrella closing about H_0. The time required for this magnetization to pass (or the umbrella to close) through one radian is called T_1. Since we cannot excite the signal again until it has returned to essentially its equilibrium value, we must generally wait three to five times T_1 between pulses. Thus T_1 is related to how often we can pulse.

Of course, we need not excite our nuclei all the way to a 90° pulse, since there will be a component in the x–y plane regardless of the size of the pulse flip angle as shown in Figure 5.6, but the intensity of that component will decrease with flip angle. Thus, while we can pulse more often, and obtain more scans per unit time, we must obtain more scans to average the weaker signals thus obtained. This situation has been analyzed by Christensen et al.[4] who have concluded that there is a large area of flip angles and pulse repetition rates that produce at least 90% of the theoretical efficiency.

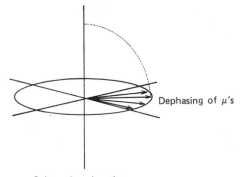

Spin–spin relaxation

Figure 5.7 Effects of spin–lattice relaxation. The various nuclear magnets do not stay in phase once the pulse is turned off and thus no longer present a strong signal to the detector; instead they begin to cancel each other out.

Typically we choose a flip angle of about 30–45° and a fairly rapid repetition rate of about $1 \times T_1$.

The other relaxation effect is spin–spin relaxation, as pictured in Figure 5.7. This amounts to a dephasing of the magnetic moments of the individual nuclei, and as they fan out, they begin to cancel each other out, leading to a less intense signal. In liquids the spin–spin relaxation time T_2 is usually less than or equal to T_1, and we find that the information in the decaying sine waves has usually died out because of T_2 effects sooner than we can pulse again because of T_1 effects. In addition, magnetic field inhomogeneity contributes to this vector dephasing. The effective T_2 that is observed as line broadening is often referred to as T_2^* to indicate that any inhomogeneities of the field are part of the estimated T_2.

CONVERTING SINE WAVES TO PEAKS

Now that we have described the fundamentals of the pulsed nmr experiment, we have determined that the result for a single line spectrum will be a single decaying sine wave as shown in Figure 5.8*a*. This is a plot of absorption of energy versus time and can thus be called a *time domain* spectrum. We can also determine by inspection that this spectrum consists of a single frequency, and we could thus make a plot of absorption versus frequency as shown in Figure 5.8*b*, indicating by that one peak that one single frequency is present.

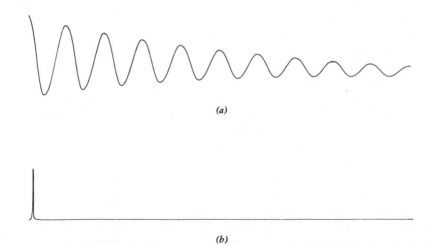

(*a*)

(*b*)

Figure 5.8   A single decaying sine wave and a plot of the single frequency it represents.

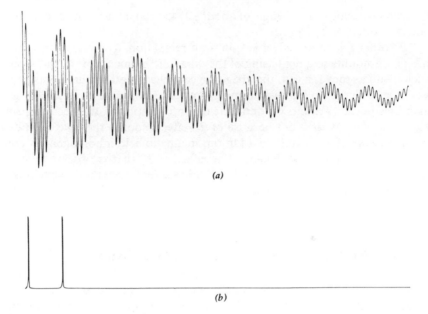

(a)

(b)

Figure 5.9   Two decaying sine waves and a plot of the two frequencies they represent.

If we observe a slightly more complex spectrum, containing two lines, as shown in Figure 5.9a, we will see two co-added sine waves decaying together. Still, by inspection, we could determine that there are two frequencies present and could again make a plot of absorption versus frequency, a *frequency domain* plot, containing two peaks. Such a plot is shown in Figure 5.9b and clearly shows two peaks representing two frequencies.

For such spectra, we do not need any sophisticated computer technology or programs, since observation of a scope trace would be sufficient. However, if we consider a spectrum such as is shown in Figure 5.10a, we see that it obviously contains a number of sine waves, but that the overall time domain spectrum is too complicated to extract all of them by inspection. Such a spectrum is called a *free-induction decay* or fid.

Figure 5.9b is a plot of absorption versus frequency for this same data, and resolves this jungle of sine waves into a few sharp lines. The time domain spectrum in Figure 5.10a was converted to the frequency domain spectrum in Figure 5.10b by using a computer program called a *Fourier transform*. For the purposes of our discussion here, we will consider this program as a "black box" that converts time domain to frequency domain spectra (and vice versa if desired). We now see the other compelling reason for using a

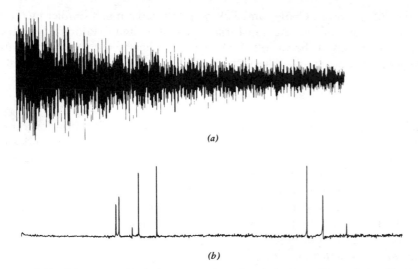

(a)

(b)

Figure 5.10 Many co-added sine waves in a free-induction decay of the ^{13}C nmr spectrum of 3-ethylpyridine (top). The Fourier transform of this free-induction decay (bottom).

minicomputer to acquire pulsed nmr data: we then have it in the computer's memory, not only to signal average if the signals are weak, but to Fourier transform it as well.

THE EQUATIONS FOR THE FOURIER TRANSFORM

The classical description of the Fourier transform amounts to a determination of each point of the transformed array by multiplying all points in the old array by an exponential function:

$$A(r) = \sum_{k=0}^{N-1} X(k) \exp\left(\frac{-2\pi irk}{N}\right), \quad r = 0 \ldots N - 1$$

where X_k are the time domain points and A_r are the frequency domain points. Excluding the method used for calculation of the exponential function, there are clearly N multiplications necessary to calculate *each* transformed point $A(r)$, and multiplications are one of the slowest functions a computer performs. The Fourier transform was, thus, for many years a computational bottleneck requiring enormous amounts of computer time.

In 1965, however, Cooley and Tukey[5] published a much simpler method for carrying out the discrete transform which amounted to taking advantage of the harmonics of the various frequencies implicit in the transform. This along with the sine look-up table used after application of Euler's formula

$$e^{iy} = \cos(y) + i \sin(y)$$

allowed a more efficient transform technique, called the Fast Fourier transform (FFT). This method required only $N \log_2(N)$ multiplications and was thus much faster. To illustrate how much faster, let us consider the case of a classical and a Cooley–Tukey transform of 16,384 points, a common array size in Fourier transform nmr. The classical method would require $(16,384)^2 = 268,435,456$ multiplications, while the Cooley–Tukey method requires only $16,384 \log_2(16,384) = 16,384(14) = 229,376$ multiplications, a savings of a factor of 1170 in time. The Cooley–Tukey method is quite simple to implement, even in a minicomputer,[6,7] and it is always used when a Fourier transform is required.

SAMPLING IN FT SPECTROSCOPY

We find that a number of things are looked at differently when we acquire Fourier transform nmr data. Most important of these is the concept of sampling *sine waves* rather than *peaks*. The data are converted to peaks later by the Fourier transform calculation, but when acquired and averaged, they are sine waves. Thus we must be concerned with how fast we need to sample in order to be sure that we have taken enough points to represent the various sine waves. It can be shown that if there are at least two points per cycle, there is only one sine wave that can be drawn through these points. Thus, if we have two or more points per cycle, we have unambiguously represented that sine wave in the computer's memory. This is similar to deciding how fine a window screen is necessary to lay over a plot of the sine wave so that two successive points will give us the shape of the sine wave.

If we have to sample two points per cycle, we must sample twice as fast as the highest frequency we have to represent. The highest frequency we can correctly observe for a given sampling rate is called the *Nyquist frequency*. For example, if we have a spectral width of 1000 Hz, then we can expect sine waves from 0–1000 Hz in the spectrum. We must then sample at a rate of at least 2000 Hz, or once every 500 μsec to ensure that we have correctly represented all frequencies. If we do not sample the data often enough, we will still see these higher frequencies, but they will seem to be the wrong ones, as we see from Figure 5.11.

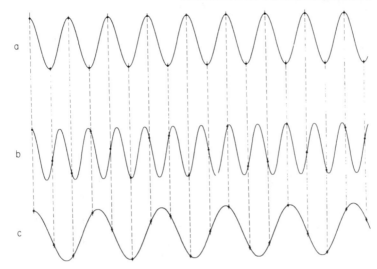

Figure 5.11   (*a*) Sampling of a sine wave at the Nyquist frequency, (*b*) sampling of frequency $N + \Delta f$, and (*c*) sampling of frequency $N - \Delta f$.

In Figure 5.11*a* we see a sine wave being sampled exactly twice per cycle, representing the highest possible frequency for that sampling rate. Figure 5.11*b* shows a frequency Δf higher than the Nyquist frequency and Figure 5.11*c* shows a frequency Δf lower than the Nyquist frequency. If we examine the amplitudes of the dots in traces *b* and *c*, we find that they are identical. Therefore the computer "sees" the higher frequency $N + \Delta f$ shown in Figure 5.11*b* as having the same frequency as $N - \Delta f$ shown in Figure 5.11*c*. This phenomenon is known as *foldback* or *aliasing* and is a problem peculiar to time domain sampling.

We can see the results of foldback in Figure 5.12, where the lowest spectrum represents the correct frequencies of the data and the upper traces are obtained by lowering the H_1 frequency by about 10 % of the spectral width each time. We see that the highest frequency line "folds back" and in this case also points downward because its phase differs. As we move the radio pulsing frequency down further, we see that the upper line eventually passes the lower line and that the upper line also eventually folds back, giving the mistaken impression that the chemical shift between the lines is different from its actual value. The phase of the lines in this example was held constant in all five traces, although computer software might ordinarily be used to try to minimize this error, and lead to even more puzzling results.

Thus we conclude that unless we are willing to accept foldback in a spectrum and are readily able to identify those aliased lines, we must accept the

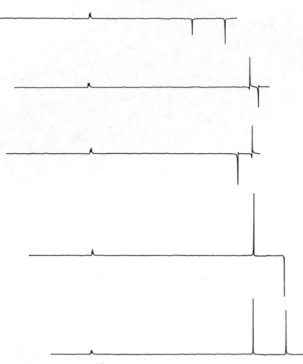

Figure 5.12   Foldback of lines in ^{13}C nmr spectrum of ethyl iodide. Bottom: the correct spectrum. In the upper traces, the pulse excitation frequency was gradually shifted downfield while maintaining the same sampling rate, leading to foldback of the upfield lines. Note that in the upper trace they have actually interchanged positions.

entire nmr spectrum and adjust our sampling rate so that we sample all the lines in the entire spectral width.

RESOLUTION IN FT-NMR

Because we are sampling in the time domain and converting to the frequency domain later by using the Fourier transform, the concepts of resolution and spectral width are reversed from those in swept nmr. In swept nmr, the resolution is determined (within the limits of the instrument) by the sweep rate: the slower the sweep, the higher the resolution, and the spectral width is determined by how long you sweep. The longer you sweep, the more frequencies you cover.

In FT-nmr, on the other hand, the spectral width is determined by the sampling rate: the faster you sample, the higher frequency sine wave you can observe, and the resolution is determined by the length of time you sample. The reason the total acquisition time determines the resolution is that the longer you sample, the more data points you have acquired into memory. Assuming that the minicomputer's memory is sufficiently large, and the spectrometer has a sufficiently high resolution magnet, the resolution is simply the reciprocal of the acquisition time, AQT.

$$R = \frac{1}{AQT} \qquad (5.8)$$

It would appear, then, that we can get more and more resolution from a given spectrum by simply sampling longer and longer. This is illustrated in Figure 5.13. This figure shows the effect of sampling the ^{13}C free-induction decay of decanol every 370 μsec (leading to a spectral width of 1350 Hz) for 0.75, 1.5, 3.0, and 6.0 sec leading to 2048, 4096, 8192, and 16,384 data points. These are Fourier transformed to the frequency domain spectra shown in the figure. While it is clear that the resolution and the S/N improve between the 2048- and the 4096-point spectrum, the S/N begins to degrade between the 4096 and the 8192 point spectrum and is substantially degraded by the 16,384-point case. This happens because the resolution corresponding to 6 sec of sampling is 0.083 Hz, and the lines in this particular sample with this particular magnet homogeneity are simply not that narrow. Consequently,

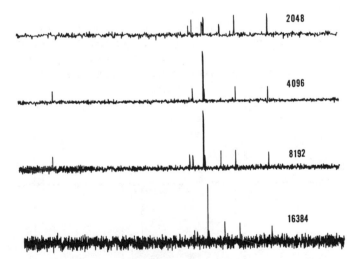

Figure 5.13 Fourier transforms of 2048, 4096, 8192, and all 16,384 points of a free-induction decay, showing that few points lead to degraded resolution (2048) and too many points lead to excess noise (16,384).

sampling that long only results in the sampling of large amounts of noise once the sine waves have died out completely. This noise is of all frequencies and will thus be spread throughout the spectrum after the transform, markedly degrading the S/N. In fact it is preferable to zero the last part of the data if it is all noise and thus decrease the noise of the transformed spectrum.

ZERO FILLING IN FT SPECTROSCOPY

In most minicomputers, the Fourier transform is done *in place*, so that the actual time domain spectrum is converted to the frequency domain data in the same array. Since the Fourier transform deals with complex numbers even if the input data is all real, the output data will consist of real and complex coefficients. Thus if we transform a 8192-point array, we will obtain 4096 words of real data and 4096 words of imaginary coefficients. The spectrum appears in both sets, but one is 90° out of phase from the other, as shown in Figure 5.14.

Since we have only half as many data points in the frequency domain, because the number of real points is only half as great, we have actually lost resolution. To gain the maximum possible resolution, we must Fourier transform our N-point fid with additional N points of zeroes. This is called *zero-filling* and it is an important technique for obtaining all the resolution that the spectrum really contains. Unfortunately, this technique requires twice as much memory as the original data, and this may not be available in a spectrometer's minicomputer. This can sometimes be solved by adding a disk for storage and processing the data in more complex ways so that the disk

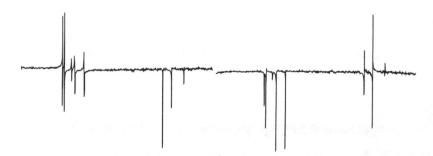

Figure 5.14   The real and imaginary parts of a Fourier transformed real free-induction decay before phase correction.

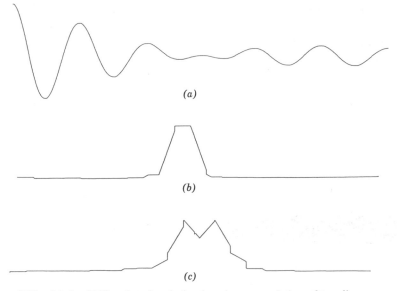

Figure 5.15 (*a*) An 8192-point, free-induction decay consisting of two lines separated by 0.12 Hz. (*b*) Expansion of the Fourier transform of (*a*), illustrating the fact that even though the data have been sampled long enough, the peaks are not resolved. (*c*) Transform of the data in (*a*) with an additional 8192 points of zeroes added to the end of the decay before phase correction.

actually is used to emulate the main memory of the computer. This is illustrated in Figure 5.15, where the transform of the same two co-added sine waves is shown with and without zero-filling.[8]

RESOLUTION VERSUS SIGNAL-TO-NOISE

The longer we sample a fid, assuming that real data persists, the more resolution information we obtain. Equation 5.8 shows that the resolution is related to the reciprocal of the acquisition time. The fid is dying out at the end of the sampling interval and only noise may remain behind. In this case, we really would like to zero out the " tail " of the fid, leaving behind only the portion containing real data. If we zero some block of points near the end of the decay, we will have introduced a step into an otherwise smooth function, and this step will transform giving odd line shapes and harmonics, depending on how large the step is.

If instead we simulate the behavior of the nuclei by multiplying the fid by a decaying exponential function, we will not introduce any steps in the data, and will affect the initial portions only negligibly, and the final portions most strongly. This will have the effect of zeroing the last points and making the intermediate ones much smaller. Since the Fourier transform of an exponential is a Lorentzian line, this could broaden any narrower lines, but there will be no noticeable effect on the wider ones. In the usual case, the weight of the exponential function is selected as a line broadening function and the FT spectrometer user simply selects a line width smaller than that of any expected lines. The effort of the function is shown in Figure 5.16: it improves the S/N by trimming out the end of the fid, but, in turn, broadens the lines slightly.

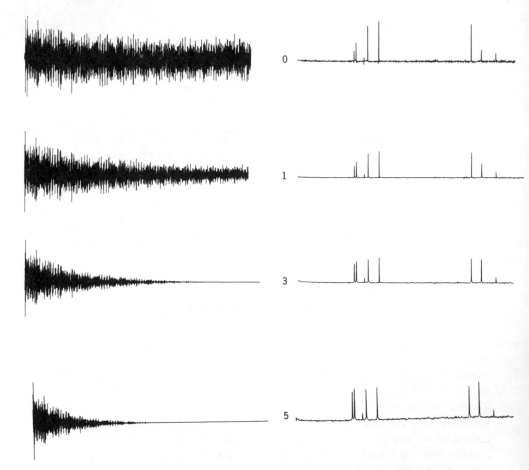

Figure 5.16 Effect of applying increasing negative exponential weighting functions on the free-induction decay and on the resulting transformed spectra. Note that while the S/N improves with an increased weighting, the line width also increases.

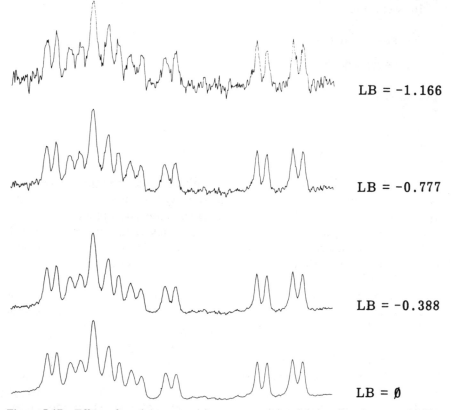

LB = -1.166

LB = -0.777

LB = -0.388

LB = 0

Figure 5.17 Effect of applying a positive exponential weighting function on a highly coupled spectrum. The S/N becomes worse, but the fine splittings (resolution) improve.

In some cases, it may also be desirable to extract more resolution information for the fid than would normally be present by enhancing the last parts of the fid. Remember: the longer you sample, the more resolution you get. The multiplication used is usually a positive exponential that leaves the beginning of the fid unchanged but inflates the end of it. The effect of this function is shown in Figure 5.17, where more resolution is apparent, but at the expense of S/N.[9]

SUMMARY OF DIFFERENCES BETWEEN SWEPT- AND PULSED-FT NMR

Let us summarize the differences in approach between data acquisition in continuous wave (cw) or swept nmr and pulsed-FT nmr.

In swept nmr:

The sweep rate determines the resolution.

The sweep time determines the spectral width.

In pulsed-FT nmr:

The sampling rate determines the spectral width.

The sampling time determines the resolution.

We can make these distinctions clear with a couple of examples. In swept (cw) nmr we find that our resolution improves if we sweep through the spectrum more slowly. Thus a 1000-sec scan usually shows better resolution than a 250-sec scan. Furthermore, the frequency range or spectral width that we scan is determined by how long we scan. Thus, if we scan for only 800 sec at 1 Hz/sec, we have scanned only 800 Hz, rather than the full 1000-Hz spectral width.

In pulsed-FT nmr, on the other hand, acquisition of data at a rate of 2000 points/sec (or 2000 Hz) leads to a spectral width equal to half the sampling rate or 1000 Hz, since we must sample twice per cycle of the highest frequency sine wave we wish to observe. If we sample at this rate for 10 sec, this will give us a resolution of 0.1 Hz. Note, however, that we cannot get this much resolution "free." We first would need 10 sec/$(5 \times 10^{-4}$ sec/point) = 20,000 memory locations in which we could store data in order to achieve this resolution. Furthermore, if we sample this long even though the free-induction decay has died out earlier, we are only adding additional noise to the data, which we will have to filter out using an exponential weighting function or by zeroing part of this excess data, thus reducing the desired resolution to the actual resolution achievable by the spectrometer and sample.

PHASE CORRECTION

When a free-induction decay is first transformed, the peaks in the frequency spectrum may not all point upward as we expect. Instead some may first appear as derivative curves and others may point downward. These peaks are said to be out of *phase*. This merely means that the peaks do not have the correct phase with respect to some reference, in this case our own expectations of peaks all pointing upward. This phase can easily be adjusted with respect to a reference frequency in cw nmr and the spectrum replotted. However, the spectrum obtained in the time domain by pulsed FT-nmr does not show easily recognizable phase information. Instead, we find that it is easier to correct the phase after the Fourier transform, since all the phase information we need is in either the real or imaginary part of the frequency domain data. This can be done digitally by multiplying the real and imaginary points by the sine and cosine of some angle, selected to make all the peaks point upward.

It is worthwhile, however, to examine the reasons for this phase error. We find that there are three major causes of these errors:

1. Spectrometer phase detector setting
2. Delay between pulse and start of data acquisition
3. Filter settings

The spectrometer phase detector is usually optimized at the beginning of a series of spectra, using a strong sample containing sharp peaks. Just as in cw spectroscopy, the smallest changes will affect this parameter: variation in sample tubes, solvents, or even spinning rate may affect the phase of the information entering the detector. This effect is *zero* order, which causes the same shift in phase for each peak regardless of frequency. Examples of out of phase and in phase cosine waves are shown in Figure 5.18 along with their transforms.

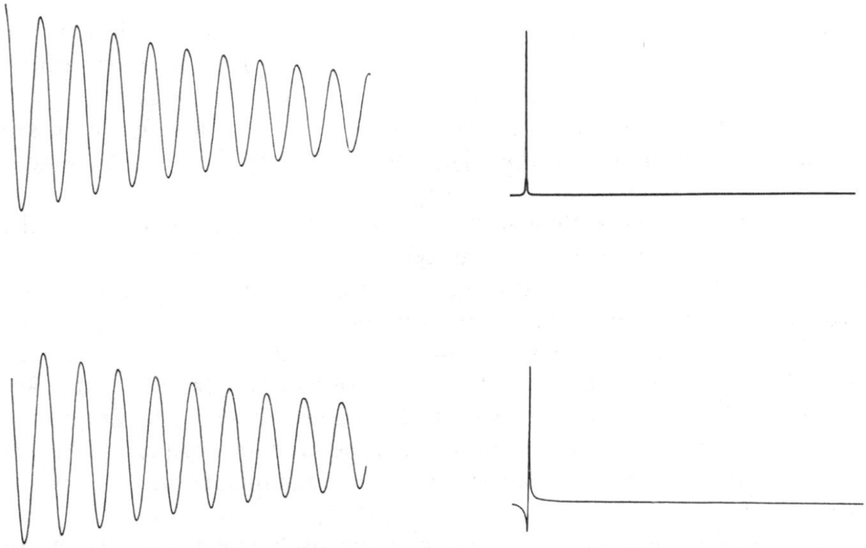

Figure 5.18   An in-phase and out-of-phase cosine wave and their transforms.

The remaining two factors, delay time and filters, have first-order effects on the spectrum. That is, at frequency *zero*, the phase shift is zero, and at the highest frequency in the spectrum, the phase shift is large. It is customary to refer to this first-order phase shift in terms of the phase angle shift of the highest frequency. A 170° first-order phase shift is one in which the first frequency domain point has zero phase shift and the last one 170° of phase shift.

The delay between the rf pulse and the time of the first data point causes a frequency-dependent phase shift related to the dwell time. The highest frequency point will be shifted by an amount equal to 180° × delay/dwell. In other words, for each dwell time unit of delay there will be a first-order phase shift of 180°. Note that in the case of first-order shifts, a 0° and a 360° shift are *not* equivalent. In the case of the 360° shift, the highest frequency point is "wrapped" around 360° in phase from the first point, causing the phases of the data points to spiral in the phase plane around one complete circle.

Logically, one can also see that first-order phase shifts can be *greater* than 360°, and can, in fact, attain almost any number of degrees. For instance, if the dwell time in a particular experiment is 30 μsec and the delay time 300 μsec, then the first-order phase shift will be

$$\frac{300 \ \mu\text{sec delay}}{30 \ \mu\text{sec dwell}} \times 180° = 1800°.$$

This is equivalent to a spiraling of the phase information in five complete circles.

Let us now consider a physical explanation for this phase shift. While the zero-order shift is quite easy to explain in terms of phase detector settings, the first-order shift is a little harder to grasp. We will first return to our discussion of sampling. Recall that the sine waves comprising the fid are sampled at a constant interval called the dwell time as shown in Figure 5.19a.

We have assumed that these waves are sampled starting exactly at 0° as the first point of the sine or cosine wave. Let us now assume that a delay of one dwell time is introduced for instrumental reasons, such as to minimize pulse feed-through. This is shown in Figure 5.19b. As you can see, the omission of the first data point causes the phase of this line to begin at about 75° instead of at 0°. This means that this line when transformed to the frequency domain will be 75° out of phase. Let us now consider a line of much higher frequency, one at the *Nyquist frequency*, or at one-half the sampling frequency. At this frequency, there are only two data points per cycle of the sine wave, as shown in Figure 5.20.

If, at the high frequency, we introduce a one dwell time delay, as shown in Figure 5.20b, we will have wiped out 180° of phase information, so that the

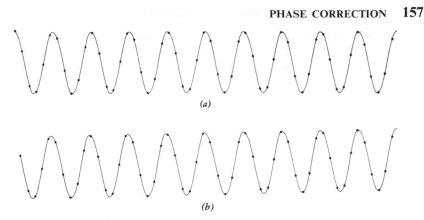

(a)

(b)

Figure 5.19 Sampling a low-frequency cosine wave with (*a*) no delay and (*b*) one-sampling interval delay. The second wave is about 75° out of phase from the first.

wave will start at 180° instead of at 0°. Thus at this frequency, the highest one represented in the fid, the delay of one single dwell time (or of one *address*) will cause a 180° phase shift. It follows from this that a two address delay would cause a 360° phase shift, and so forth.

A line that is " folded back " or aliased from a frequency higher than twice the sampling rate will have the first-order phase dependence that it would have if it were actually observed at its correct frequency. Thus a folded back line will seem to have an anomalous phase relative to lines around it. If a line appears to be folded back, then alter the irradiation frequency and obtain a new spectrum. If all lines move in the same direction except the suspicious one, it is folded back.

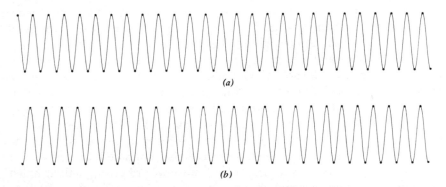

(a)

(b)

Figure 5.20 Sampling a Nyquist frequency cosine wave with (*a*) no delay and (*b*) a one-sampling interval delay. The resulting wave is 180° out of phase from the wave having no delay.

A line may also appear to have anomalous phase if its spin–spin relaxation time, T_2, is much longer than that of other lines in the spectrum and if the pulse repetition rate is fairly rapid. In this case, spin–echo effects will introduce some phase anomalies.

The third cause of phase shifts are the characteristics of the filter used to limit the spectrum bandwidth. The four-pole Butterworth filter, for example, causes a first-order shift of $-180°$ if used in the system. This does not vary appreciably with the cutoff frequency selected. Thus the total formula for determining the first-order phase shift is given by

$$\frac{\text{delay time}}{\text{dwell time}} \times 180° - 180°.$$

Phase correction is performed by convolving all points in the real and imaginary spectra with a cosine and sine wave. Zero-order phase correction is performed by

$$x'_j = x_j^{\text{real}} \cos (A) - x_j^{\text{imag}} \sin (A)$$

where j is the index of the point and A is the zero-order phase correction angle.

First-order phase correction is performed by

$$x'_j = x_j^{\text{real}} \sin (jB/N) + x_i^{\text{imag}} \cos (jB/N)$$

where N is the total number of frequency domain points and B is the first-order phase correction constant.

QUADRATURE DETECTION

Up to this point we have discussed nmr as if the frequency of the pulse was at one end of the spectrum and the data were all above this frequency. This was the common way to do the experiment for some years. However, more recently it has been recognized that the spectral bandwidth is actually *plus or minus* one-half the sampling frequency, and that if the radio frequency is placed at one end, the noise from the same spectral width below the pulse frequency folds back into the spectrum and decreases the S/N as shown in Figure 5.21. If, instead, we could place the radio frequency of the carrier in the middle of the spectrum and somehow distinguish positive from negative frequencies, we will be able to obtain more S/N per unit time, since all the noise on the " other " side of the carrier would be eliminated. In fact, we will be able to improve S/N by a factor of $\sqrt{2}$ since we are getting half as much noise or twice as much signal per unit time.

We can, in fact, distinguish positive frequencies from negative ones if we use two phase-sensitive detectors whose phase is set $90°$ apart. These detectors

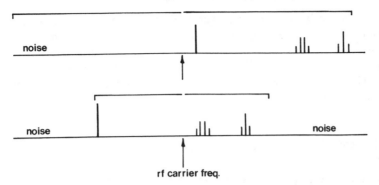

noise

noise   noise

rf carrier freq.

Figure 5.21 Top: single detector FT-nmr—the rf pulse frequency is placed at one end of the spectrum. Noise folds back from the other side of the carrier. Bottom: quadrature detection FT-nmr—the rf carrier is placed in the center of the spectrum and the spectral width is decreased by half. Noise cannot fold back from within the spectral width—resulting in a greater S/N.

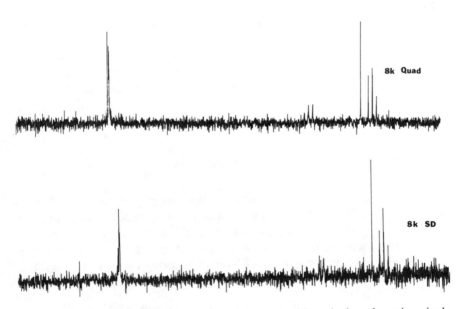

8k Quad

8k SD

Figure 5.22 The same FT nmr spectrum obtained with a single pulse using single detector and quadrature detection methods. Note that the S/N improvement is about 1.4 as predicted.

159

are then said to be in *quadrature* and their resulting data are channeled into adjacent arrays. These data then represent the real and imaginary input to a complex Fourier transform and the resulting frequency position of the carrier. The same spectrum as a single scan in single detector and quadrature detector nmr is shown in Figure 5.22. Note the improvement in S/N is about 1.4 as predicted.

DECOUPLING IN PULSED-FT SPECTROSCOPY

For a time it was popularly assumed that decoupling of one line from another in a proton spectrum was definitely a task more suited to swept nmr, and that this could not be done conveniently by FT techniques. In fact, the decoupling task is even easier than ever with current spectrometers, since the user need only run the normal spectrum and pick out the peak he wishes to irradiate using an intensified dot or cursor on the spectrometer-computer's scope display. Then the entire decoupling experiment can be run immediately without the troublesome plotting of the spectrum and careful calibration of the decoupling frequency.

Decoupling is instrumentally more involved, but it is a standard feature of all instruments these days. It is accomplished by setting a second oscillator to the desired decoupling frequency and then turning it on at low power between acquired data points. At this low power setting only the single frequency is affected and the receiver never " sees " this frequency since it is turned off during the time a data point is taken. The decoupled spectra of crotonaldehyde shown in Fig. 4.24 (see p. 99) were acquired in this way.

CORRELATION SPECTROSCOPY

The goal of rapid acquisition of data by signal averaging can be met by a technique other than pulsed-FT spectroscopy. We have pointed out that we are limited to sweeping at about 1 Hz/sec in cw nmr so that nearby lines do not oscillate into each other as we begin to pass out of one resonance and into the next. There is a method, however, known as *correlation spectroscopy*, for acquiring data rapidly by fast scanning and then converting it to a conventional spectrum later.

To utilize this technique, pioneered by Dadok and Sprecher,[10] we sweep very rapidly through a region of interest, say in 5–30 sec rather than 100–500 sec/scan, and signal average the resulting data into a computer's memory as before. We then *cross correlate* these data with that obtained by scanning rapidly through a single, sharp line such as TMS at the same rate and through

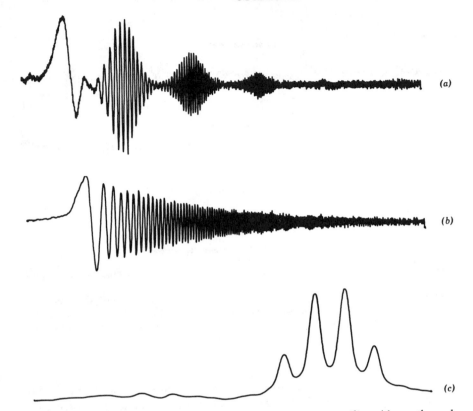

Figure 5.23   (*a*) Rapid scan through the quartet of ethylbenzene; (*b*) rapid scan through the TMS singlet under the same sweep conditions; (*c*) cross-correlation product of (*a*) and (*b*).

the same sweep width. The result of this correlation is a conventional spectrum as shown in Figure 5.23.

A more versatile technique is to perform the cross correlation between the acquired data and the theoretical response of an infinitely sharp line under the same conditions.[11,12] This adds more resolution to the data since in this case there is no added line width contribution from the theoretical line. A typical experimental result is shown in Figure 5.24 for phenylethanol.

Cross correlation, if carried out discretely, is the product of every point in the experimental spectrum with every point in the reference spectrum:

$$z(k) = \sum_{i=0}^{N-1} h(i) \cdot x(k + i) \qquad k = 0 \cdots N - 1.$$

where $x(i)$ is the experimental spectrum and $h(i)$ the reference spectrum.

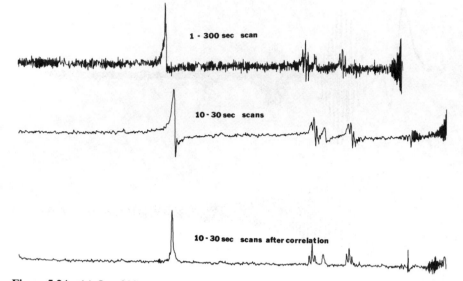

Figure 5.24 (*a*) One 300-sec scan through the proton nmr spectrum of phenylethanol. (*b*) Ten 30-sec scans through the same spectrum; (*c*) Spectrum (*b*) after correlation with a theoretical function. Note the improved S/N and corrected phase that is possible in the same acquisition time.

This amounts to N^2 multiplications and can take an extremely long time to carry out. By using Fourier transform techniques, however, the correlation process can be reduced to two transforms and N multiplications. First, the discrete FT of the data is computed:

$$X(n) = \sum_{k=0}^{N-1} x(k) \exp\left(\frac{-2\pi i n k}{N}\right) \qquad n = 0 \cdots N - 1$$

and of the reference

$$H(n) = \sum_{k=0}^{N-1} h(k) \exp\left(\frac{-2\pi i n k}{N}\right) \qquad n = 0 \cdots N - 1.$$

Then the product of the two arrays is calculated:

$$Y(n) = X(n)H(n) \qquad n = 0 \cdots N - 1$$

and the inverse FT of the product is computed

$$y(k) = (1/N) \sum_{i=0}^{N-1} Y(n) \exp\left(\frac{2\pi i n k}{N}\right).$$

This process amounts to only $N = 3 \log_2 N$ multiplications rather than the N^2 multiplications required by the direct method and is thus substantially faster for large arrays.

The correlation technique is in many ways complementary to the pulsed FT technique since it can be accomplished (a) without high powered pulse equipment and (b) over any desired spectral width. Thus while very large solvent peak signals would cause pulsed-FT data to fill memory quite rapidly, the correlation nmr technique allows the user to scan any desired spectral range, thus eliminating the solvent peak if there are not other peaks of interest nearby.

PROBLEMS

Problem 5.1

How fast must the computer sample data in order to represent a spectral width of 4200 Hz?

Problem 5.2

A scientist plans to observe a wide range of ^{31}P chemical shifts. If the system in use observes protons at 200 MHz and ^{31}P is seen at 36.44 MHz when protons are observed at 90 MHz, what sampling rate must the scientist use to observe 25 ppm of phosphorous spectra?

Problem 5.3

Is there any advantage to the data acquisition system when quadrature detection is installed?

Problem 5.4

Quadrature detection is often spoken of as being used to "avoid foldback." Under what conditions is this true? When is it not true?

REFERENCES

1. J. W. Cooper, *Computers and Chem.*, **1**, 55 (1976).
2. J. W. Pople, W. G. Schneider, and H. J. Bernstein, *High-Resolution Nuclear Magnetic Resonance*, McGraw-Hill, New York, 1959, p. 37.
3. R. R. Ernst and W. A. Anderson, *Rev. Sci. Instrum.*, **37**, 93 (1966).

4. K. A. Christensen, D. M. Grant, E. M. Shulman, and C. Walling, *J. Phys. Chem.*, **78**, 1971 (1974); E. D. Becker, J. A. Ferretti, and P. M. Gambir, *Anal. Chem.*, **51**, 1413 (1979).

5. J. W. Cooley and J. W. Tukey, *Math. Comput.*, **19**, 297 (1965); W. T. Cochran et al., *IEEE Trans Audio-Electroacoutics*, **AU15**, 45–55 (1967); L. R. Rabiner and C. M. Rader, *Digital Signal Processing*, IEEE Press, New York, 1972.

6. J. W. Cooper, " Data Handling in Fourier Transform Spectroscopy," in *Transform Techniques in Chemistry*, P. Griffiths, Ed., Plenum Press, New York, 1978.

7. J. W. Cooper, *The Minicomputer in the Laboratory*, Wiley-Interscience, New York, 1977.

8. J. W. Cooper, *Anal. Chem.*, **50**, 801A, (1978).

9. A. O. Clouse et. al., *J. Am. Chem. Soc.*, **95**, 2496 (1973).

10. J. Dadok and R. F. Sprecher, *J. Magn. Resonance*, **13**, 243 (1974).

11. R. K. Gupta, J. A. Ferretti, and E. D. Becker, *J. Magn. Resonance*, **13**, 275 (1974).

12. J. W. Cooper, " Advanced Techniques in Fourier Transform Nmr," in *Transform Techniques in Chemistry*, P. Griffiths, Ed., Plenum Press, New York, 1978.

CHAPTER SIX

^{13}C NUCLEAR MAGNETIC RESONANCE SPECTROSCOPY

Now that we have seen how we can signal average weak signals such as we might find in dilute samples by FT spectroscopy, we can appreciate the nature of the problem in obtaining ^{13}C spectra. Such spectra are sometimes termed *cmr* spectra as contrasted with ordinary *pmr* (proton) spectra. ^{12}C has a nuclear spin of 0 but ^{13}C has a spin of 1/2. However, the ^{13}C isotope is a so-called "rare" spin and has a natural abundance of only 1.1%. The sensitivity problem this creates is overcome by several techniques:

1. Use of signal averaging
2. Rapid acquisition by FT spectroscopy
3. Use of 10 mm (or larger) sample tubes
4. Decoupling of proton spins from carbon spins
5. Utilization of the nuclear Overhauser effect (NOE)

COUPLING IN ^{13}C SPECTRA

Since ^{13}C has a relatively low natural abundance, the spectrum of an un-enriched compound can be regarded as a spectrum of a mixture of molecules, each containing only one ^{13}C atom. A simple four-line spectrum like that of 2-bromobutane shown in Figure 6.1 is really the spectrum of four separate groups of molecules, each containing a ^{13}C in a different position. This means that we seldom see very much coupling between adjacent carbons in un-enriched samples. We do, however, see substantial coupling between the carbons and their attached hydrogens, and indeed with more distant hydrogens in many cases. For this reason, the proton coupled spectra of organic molecules are quite complex.

Rather than decoupling each proton line from the carbon spectrum individually as is common in the decoupling of proton spectra, the entire range of ^{1}H chemical shifts is irradiated with all frequencies in the proton chemical

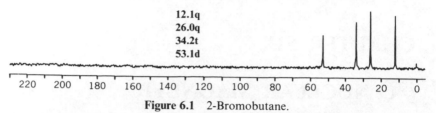

Figure 6.1 2-Bromobutane.

shift range in the form of *white noise* so that all of the proton lines are decoupled at once. The coupled and decoupled spectra of ethylbenzene are shown in Figure 6.2.

We notice immediately that there is a substantial difference in S/N in these spectra, obtained with the same number of scans. This occurs principally because the multiplets from the ^{13}C—H coupling collapse, but also because of the NOE.

THE NUCLEAR OVERHAUSER EFFECT

If we were to integrate the intensities in Figure 6.2a and compare them with Figure 6.2b, we would find that the areas in Figure 6.2a are greater than those in Figure 6.2b by factors up to 2.988. This effect is caused by the decoupling of the hydrogen spins from the carbon spins which leads to a transfer of spin polarization to the carbons and a consequent increase in the carbon intensities. This is one of a number of effects discovered by Overhauser and is termed the *nuclear Overhauser effect* or NOE. The explanation below follows that of Noggle and Schirmer.[1]

To understand why this effect occurs, we will take a simple case of two spins, labeled C and H, which form an AX system. We need not require that they have a coupling constant J, but only that they are sufficiently close that one can contribute to the relaxation of the other. Two spins can have four possible spin states: $\alpha\alpha$, $\alpha\beta$, $\beta\alpha$, and $\beta\beta$. We will make the assumption that the β state is the lower one and recall that allowed transitions can only occur which cause a change in total spin of ± 1. Thus transitions from $\alpha\alpha$ to $\alpha\beta$ are allowed directly, but not $\alpha\alpha$ to $\beta\beta$.

In the energy level diagram in Figure 6.3 we see that there are four transitions which are allowed and that they have only two energies, W^H and W^C. The two quantum transition W^2 is not allowed and occurs only indirectly as a relaxation effect. At equilibrium, there will be only a slight difference in the population of these energy levels so that levels 2 and 3 will have a population B, the lower level $B + \delta$, and the upper level $B - \delta$. The intensities of the nmr lines will be proportional to differences in population between levels

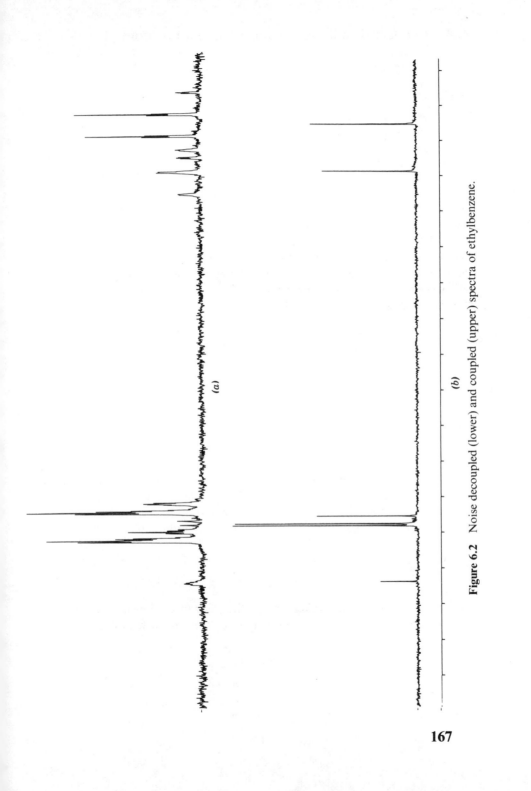

Figure 6.2 Noise decoupled (lower) and coupled (upper) spectra of ethylbenzene.

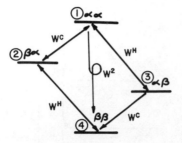

Figure 6.3 Transitions of the proton and ¹³C spins and their relaxation through a "forbidden" pathway W^2 because of relaxation coupling.

where transitions can occur. If we represent the population of each level i by the symbol P_i, then the intensity of the transition W^C will be proportional to the differences $P_2 - P_1$ and $P_4 - P_3$. If there is no J-coupling between the spins, these transitions will be of equal energy and only one line will result. Similarly, if we observe the proton spectrum, we can see a single line due to W^H whose intensity will be proportional to $P_3 - P_1$ and $P_4 - P_2$. This is summarized below:

Level	Equilibrium populations
1	$B - \delta$
2	B
3	B
4	$B + \delta$

Carbon transition intensities: $P_2 - P_1 = \delta$
$$P_4 - P_3 = \delta$$

Let us now saturate the H spins by applying a high power decoupling field to the sample. Saturation means that the probability of an upward transition is exactly equal to the probability of a downward transition by relaxation and that the levels have exactly equal populations. Thus we have made the populations of levels 1 and 3 and of 2 and 4 equal:

$$P_1 = P_3$$

and

$$P_2 = P_4.$$

When we do this we have taken half of the spins in the lower level and promoted them to the upper level. Our spin populations now look like this:

Level	Saturated H
1	$B - \delta/2$
2	$B + \delta/2$
3	$B - \delta/2$
4	$B + \delta/2$

Carbon transition intensities: $P_2 - P_1 = \delta$
$$P_4 - P_3 = \delta$$

If this were all that happened and no relaxation occurred between the two spins, there would be no observed NOE. However, the population difference between levels 1 and 4, $P_1 - P_4$, is equal to 2δ at equilibrium, but is only equal to δ when H is saturated. The relaxation pathway W^2 tends to try to restore this equilibrium population difference by relaxing spins into level 4 from level 1. If we represent the number of spins so relaxed by d, we then have the following populations:

Level	Saturated H $+$ W^2 relaxation
1	$B - \delta/ -d$
2	$B + \delta/2$
3	$B - \delta/2$
4	$B + \delta/2 + d$

Carbon transition intensities: $P_2 - P_1 = \delta + d$
$$P_4 - P_3 = \delta + d$$

We now see that the intensity of each of the W^C lines is greater by an amount d because of this relaxation. This intensity increase has a theoretical maximum of 2.988, but may be less because of less efficient relaxation and because of another relaxation pathway between levels 3 and 2 which tends to decrease the NOE. This NOE effect has been illustrated here for uncoupled systems that share a relaxation pathway, but it has been shown that it holds for loosely coupled systems as well, where $J_{AX} \ll (v_A - v_X)$, which is certainly always true for coupling between different kinds of nuclei. Thus we see that our carbon spectral lines will have an added S/N enhancement of nearly 3 because of the NOE effect and will also increase in intensity because of the collapse of complex multiplets.

OFF-RESONANCE DECOUPLING

Since ^{13}C spectra are commonly simplified and intensified by broadband proton noise decoupling, there is less information present about the various lines than in coupled proton spectra. While carbon–carbon couplings are fairly rare in routine spectra, it is clear that the carbon–hydrogen couplings would give a substantial amount of information about the number of hydrogens connected to a given carbon. These couplings can be very complex, however, and seldom produce simple first-order spectra.

To simplify these spectra, the technique of off-resonance decoupling is employed. This technique utilizes a single frequency source of high power

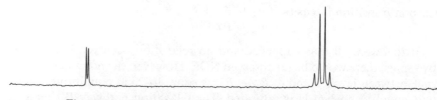

Figure 6.4 Off-resonance decoupled spectrum of isopropanol.

irradiating at one end of the proton spectral region. The effect of this irradiation is the simplification of the carbon spectra by the removal of all long-range couplings, leaving only simple multiplets in the carbon spectrum corresponding to the number of attached hydrogens. This is illustrated in Figure 6.4 for isopropanol. Note the quartet for the methyl carbon and the doublet for the methine carbon.

Since only a single frequency is used for off-resonance decoupling, the hydrogens having chemical shifts closer to the irradiation frequency will be decoupled most effectively and those furthest away least effectively. This information can sometimes be used to determine structural information, since the proton frequency of an attached proton may give structural information about a particular carbon atom. It is common to select a proton irradiation frequency upfield of all resonances so that carbons attached to downfield protons will show multiplets with larger apparent couplings.

GATED DECOUPLING

By turning the proton broad band decoupler on and off during ^{13}C data acquisition, it is possible to obtain two very useful effects: coupled spectra with NOE enhancement and decoupled spectra without NOE enhancement.

If the ^{13}C signal is acquired with the decoupler on during data acquisition but off between the time sampling ends and the beginning of the next pulse, the resulting spectrum will be decoupled, but the NOE will not have time to build up. This occurs because decoupling is an instantaneous phenomenon involving saturating the ^{1}H levels, but the NOE will build up only after the ^{1}H—^{13}C relaxation begins to take effect. These sequences are illustrated in Figure 6.5.

While such a spectrum will in and of itself show less S/N than a conventional noise-decoupled ^{13}C spectrum, it is an invaluable aid in measuring the NOE. NOE measurement is accomplished by obtaining the conventional decoupled spectrum and the gated-decoupled spectrum and dividing one

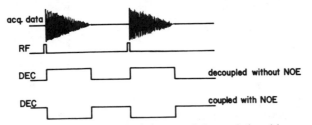

acq. data

RF

DEC ⎍⎍ decoupled without NOE

DEC ⎍⎍ coupled with NOE

Figure 6.5 The timing necessary for obtaining gated decoupled and inverse-gated decoupled spectra.

set of peak areas by the other. These are illustrated in Figure 6.6. The value of the NOE can be a useful tool in line assignment in complex spectra. While we certainly expect a small NOE for all quaternary carbons, the distance to the nearest hydrogen will determine the observed NOE. Thus quaternary carbons deeply buried inside complex structures will have even less NOE than those on the outside of the molecule.

The converse experiment, in which the decoupler is off during ^{13}C data acquisition but on during the interpulse delay, is used to obtain coupled

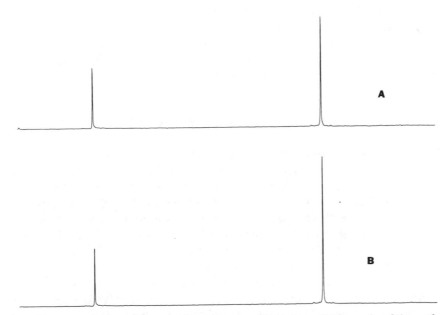

A

B

Figure 6.6 Gated and noise decoupled spectra of isopropanol. The ratio of the peaks in the lower (noise decoupled) trace to those in the upper (gated decoupled) trace allows one to determine the NOE for each peak in the spectrum.

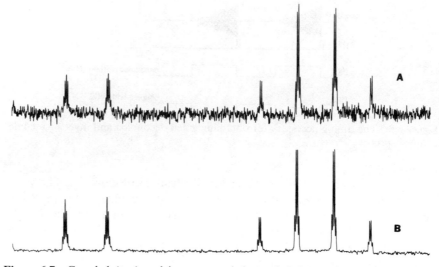

Figure 6.7 Coupled (top) and inverse-gated decoupled (bottom) spectra of isopropanol (same number of scans).

spectra with NOE-enhanced intensities as shown in Figure 6.7. For this experiment, the delay time between pulses must be relatively long so that the NOE has time to build up before the pulse is given for the next scan. This effect is possible because while the decoupling disappears instantaneously, the NOE decays with a time constant close to T_1.

SAMPLE PREPARATION FOR ^{13}C SPECTROSCOPY

In most modern ^{13}C spectrometers the sample tube diameter is at least 10 mm and some spectrometer magnets and probes will accommodate a 15 mm, 20 mm, or even 40 mm sample size. The large tubes must be carefully constructed so that they can be spun without wobbling and are thus quite expensive. Furthermore, these larger tubes very easily develop a vortex when spun and the liquid is thus topped with a plastic vortex plug that keeps the liquid held in position and does not allow the vortex to form.

Since the nuclei to be studied in a ^{13}C nmr experiment are not highly abundant, they do not present a suitable lock signal and locking is generally on another nucleus. Since the proton signals are commonly decoupled, ^{1}H cannot be used for locking and the signal from the deuterium in deuterated solvents is commonly used. Other routine spectrometers employ an external lock, which may be ^{19}F or deuterium.

The solvent used in ^{13}C nmr is thus nearly always deuterated and almost always contains carbons as well. The solvent therefore usually produces a signal that is observed in the nmr spectrum and these solvent lines must be recognized and disregarded in analyzing the spectrum. Some of the common solvents are:

Solvent	Chemical Shift (multiplicity)	
Acetone-d_6	δ206.0(13)	29.8(7)
Benzene-d_6	128.0(3)	
Chloroform-d	77.0(3)	
Cyclohexane-d_{12}	26.4(5)	
Diethyl ether-d_{10}	65.3(5)	14.5(7)
Dimethyl sulfoxide-d_6	39.5(7)	
Methylene chloride-d_2	53.8(5)	
Methanol-d_4	49.0(7)	
THF-d_8	67.4(5)	25.3(br)

Note that none of the solvent lines are singlets, since they are coupled to deuterium which is *not* being decoupled from the carbons. Recall that the splitting caused by the coupling of one nucleus to another is given by $2nI + 1$, where I is the spin of the coupled nucleus. The spin of deuterium is 1, and the multiplicity of the carbon in CDCl$_3$ is thus 3 and in CD$_3$OD is 7.

Often, however, the solvent lines are not as prominent as their concentration might predict, since many of them have extremely long T_1s. Thus when

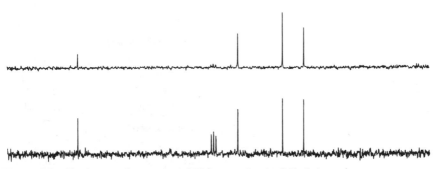

Figure 6.8 Single scan (bottom) and 100 scans (top) of diethyl succinate,

$$CH_3CH_2OOCCH_2CH_2COOCH_2CH_3$$

in CDCl$_3$, illustrating the reduced intensity of the CDCl$_3$ and carbonyl lines which have T_1's much longer than the chosen pulse repetition rate.

the sample is pulsed fairly rapidly with, say 30–45° pulses, the solvent line will not have relaxed very much and will produce only a rather weak signal. This is illustrated in Figure 6.8. The downfield carbonyl carbon is also reduced in intensity in this spectrum by rapid pulsing since carbonyl carbons also have rather long T_1s.

CHEMICAL SHIFTS IN ¹³C SPECTRA

Unlike proton spectroscopy, most ¹³C spectra consist of sharp resolved lines for each magnetically different carbon in the compound. These cover an extremely large chemical shift range of about 200 ppm and, like proton spectra, are measured relative to the TMS line. The carbons in tetramethylsilane are all equivalent and are highly shielded as are the protons, so that TMS makes an ideal reference for the carbon spectrum as well, coming at the extreme upfield end of the spectrum for most compounds. In a general way,

Table 6.1   Chemical Shifts of Common ¹³C Functional Groups (ppm from TMS)

Alkanes		Ethers	
Cyclopropanes	0–8	CH_3—O	45–60
Cycloalkanes	5–25	RCH_2—O	42–70
R—CH_3	5–25	R_2—CH—O	65–77
R—CH_2—R	22–45	R_3—C—O	70–83
R_2CH—R	30–58	**Unsaturated Compounds**	
R_3—C—R	28–50	Aromatics	110–133
		Alkenes	100–143
		Alkynes	75–95
Halogens		**Carbonyl Carbons**	
CH_3X	5–25	R—CO—OR	160–177
RCH_2X	5–38	R—COOH	162–183
R_2CHX	30–62	RCHO	185–205
R_3CX	35–75	R—CO—R	190–220
		Heteroatoms	
		RCH_2—S	22–42
Amines		RCH_2—P	10–25
CH_3—N	10–45	Ar—P	120–130
R—CH_2—N	45–55	Ar—N	130–138
R_2—CH—N	50–70	Ar—O	130–150
R_3—C—N	60–75	R—CN	118–123

the chemical shifts of carbons follow the same shielding and deshielding characteristics that we find in the proton spectra, with the major exception that the ring current effects observed for aromatic protons are not observed for aromatic carbons. A table of some common carbon chemical shifts is shown in Table 6.1.

ALKANE CHEMICAL SHIFTS

Aliphatic carbons are observed in the region of -2–50 ppm from TMS. The more substituted the carbon, the further downfield it is found. This is called the α-effect, and it amounts to about $+9$ ppm for each carbon added.

CH_4 CH_3—CH_3 CH_3—$\underline{C}H_2$—CH_3 CH_3—$\underline{C}H(CH_3)_2$ $(CH_3)_4\underline{C}$
$\delta-2.1$ $\delta5.9$ $\delta16.1$ $\delta25.2$ $\delta27.9$

Additional carbons substituted β to a given carbon also cause downfield shifts of about 9 ppm. This is called the β-effect and is illustrated for the terminal carbon in ethane, propane, isobutane, and neopentane below:

$\underline{C}H_3$—CH_3 $\underline{C}H_3$—CH_2—CH_3 $\underline{C}H_3$—$CH(CH_3)_2$ $\underline{C}H_3$—$C(CH_3)_3$
$\delta5.9$ $\delta15.6$ $\delta24.3$ $\delta31.5$

Note that this is an increasing downfield shift with the addition of each new carbon to the next one in the chain.

A similar effect is observed on the chemical shift of the carbon once removed from that being successively substituted and is referred to as the γ-effect. It has, however, the opposite sign and causes an upfield shift of about -2.5 ppm. This is illustrated below:

$\underline{C}H_3$—CH_2—CH_3 $\underline{C}H_3$—CH_2—CH_2—CH_3
   $\delta15.6$ $\delta13.2$

$\underline{C}H_3$—CH_2—$CH(CH_3)_2$ $\underline{C}H_3$—CH_2—C—$(CH_3)_3$
   $\delta11.5$ $\delta8.7$

These three simple facts allow one to assign the spectra of many alkanes and alkyl side chains fairly easily.

For example, let us consider the spectrum of n-heptane shown in Figure 6.9. We find only four lines at $\delta14.2$, $\delta23.1$, $\delta29.1$, and $\delta32.4$. If off-resonance decoupling information is included, we find that the upfield line is split into a quartet and the rest into triplets. Thus the line at $\delta14.2$ is due to the methyl groups and the three remaining lines to the three types of methylene carbons. Interestingly, if the intensity information is even remotely correct, we see that the least intense line is not the furthest downfield, although it is due to the C-4 CH_2 group. If we consider the α-, β-, and γ-effects described above,

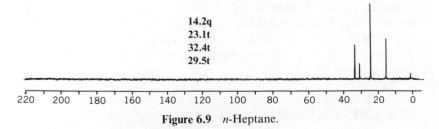

Figure 6.9 *n*-Heptane.

however, we see that this is consistent with these observations. The C-2 line is substituted by a single CH₃ group (the C-1 carbon) and is thus shifted downfield about 9 ppm by the α-effect. The C-3 carbon is shifted downfield by an additional 9 ppm since both the α- and β- effects are operating. However, the C-4 carbon is shifted *upfield* from C-3 by about 2 ppm by the γ-effect and is thus upfield of C-3. This sort of line shuffling is commonly seen in the spectra of alkanes and should be remembered whenever assignments are to be made.

EQUATIONS FOR ASSIGNING ALKANE SHIFTS

One of the great simplifications of ¹³C nmr is that the various substituent effects are nearly always additive, making the assignment of chemical shifts often a matter of simply adding up the predicted shift effects. Thus the above empirically deduced effects can be expressed mathematically, using constants that were carefully determined by regression analysis by Grant and Paul,[2] leading to the equation

$$\delta(k) = -2.1 + \sum n_{ik}A_i \qquad (6.1)$$

where A_i is the amount to add for each substituent in the ith position and n_{ik} is the number of carbons in the ith position relative to carbon k. For this equation the values of A are:

$A_1 = +9.1$ ppm (α-effect)

$A_2 = +9.4$ ppm (β-effect)

$A_3 = -2.5$ ppm (γ-effect)

For heptane, this equation predicts:

	Calculated	Experimental
$\delta_1 = -2.1 + 9.1 + 9.4 - 2.5 =$	13.9	14.2
$\delta_2 = -2.1 + 2(9.1) + 9.4 - 2.5 =$	23.0	23.1
$\delta_3 = -2.1 + 2(9.1) + 2(9.4) - 2.5 =$	32.4	32.4
$\delta_4 = -2.1 + 2(9.1) + 2(9.4) + 2(-2.5) =$	29.9	29.1

in excellent agreement with the measured values. Note that these calculation procedures, while amusing on known compounds, can help us make line assignments in unknown compounds.

For larger alkanes, both linear and branched, a somewhat more complex formula takes account of the differences between CH, CH_2, and CH_3 groups α to the carbon of interest and includes a δ-effect. The formula due to Lindeman and Adams[3] is:

$$\delta(k) = A_n + \sum_{m=0}^{2} N_m^\alpha \, \alpha_{nm} + N^\gamma \, \gamma_n + N^\delta \, \delta_n \qquad (6.2)$$

where n = the number of hydrogens at carbon k

m = the number of hydrogens attached to the α-carbon

N_m^α = the number of CH_m groups at the α-position. It varies from 0–2; α-methyl groups are ignored

N^γ = the number of γ-carbons

N^δ = the number of δ-carbons

The values for these various parameters are given in Table 6.2.

Using these parameters, let us calculate the predicted chemical shift for the C-4 of 2,2,3-trimethylbutane:

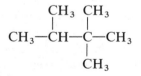

C-4 has three hydrogens, so we use only the top rows of the table for $n = 3$. There are no α-carbons having zero hydrogens attached, one having one attached and none having two attached, β carbons do not appear in the formula, and there are three γ-carbons. There are no δ-carbons. The formula is then evaluated as:

$$\delta = A_3 + 0\alpha_{30} + 1\alpha_{31} + 0\alpha_{32} + 3\gamma_3 + 0\delta_3$$
$$\delta = 6.80 + 0(25.48) + 1(17.83) + 0(9.56) + 3(-2.99) + 0(0.49) = \delta 15.66$$

Table 6.2   Parameters for Estimating Chemical Shifts by Lindeman-Adams Equation

n	A_n	m	α_{nm}	γ_n	δ_n
3	6.80	2	9.56	−2.99	0.49
		1	17.83		
		0	25.48		
2	15.34	2	9.75	−2.69	0.25
		1	16.70		
		0	21.43		
1	23.46	2	6.60	−2.07	0
		1	11.14		
		0	14.70		
0	27.77	2	2.26	0.86	0
		1	3.96		
		0	7.35		

This compares favorably with the experimental value of $\delta 15.9$.

In a similar fashion, we can predict the shift of the CH(3) carbon. This carbon has one hydrogen attached, so $A_1 = 23.46$. Attached α to this carbon are two methyl groups (which are ignored in the calculation) and one quaternary carbon. There are no γ- or δ-carbons. The predicted shift is thus calculated as:

$$\delta = 23.46 + (1)14.70 + 0(-2.07) + 0(0) = 38.16$$

which compares well with the experimental shift of $\delta 38.1$.

Quaternary carbons are usually easily recognized in alkanes; since they have no attached hydrogens, they will be much less intense. This lower intensity is due to a much lower NOE, and since such carbons have longer T_1's, rapid pulsing may partially saturate them.

ALKENES

Most alkene carbons are found in the range of 100–165 ppm. In the longer alkenes, the difference between the greater and lesser substituted of the alkene carbons is substantial, with the carbon having the shorter chain attached

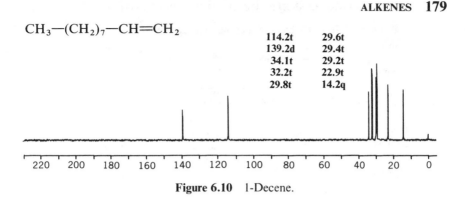

$$CH_3—(CH_2)_7—CH{=\!=}CH_2$$

114.2t	29.6t
139.2d	29.4t
34.1t	29.2t
32.2t	22.9t
29.8t	14.2q

Figure 6.10 1-Decene.

always at higher field. A terminal methylene is generally nearly 24 ppm upfield from the other olefinic carbon, and there is often a 7–10 ppm difference for 2-alkenes and a 1–2 ppm difference for 3-alkenes. For example, in the spectrum of 1-decene in Figure 6.10, we find the terminal methylene carbon at 114.2 and C-2 at 139.2.

As with alkanes, a series of additive parameters has been developed for the substitution of additional carbons at the α-, β-, and γ-positions.[4] They are shown schematically for the underlined carbon in the diagram below:

$$
\begin{array}{ccccccccc}
\overset{\delta}{C} & \!\!—\!\! & \overset{\beta}{C} & \!\!—\!\! & \overset{\alpha}{C} & \!\!—\!\! & \underline{C} \!\!=\!\! C & \!\!—\!\! & \overset{\alpha'}{C} & \!\!—\!\! & \overset{\beta'}{C} & \!\!—\!\! & \overset{\gamma'}{C} \\
-1.5 & & 7.2 & & 10.6 & & & & -7.9 & & -1.8 & & +1.5
\end{array}
\quad +123.3 \quad (6.3)
$$

Calculating the chemical shift of C-3 in 3-hexene, we would find one each of α-, β-, α'-, and β'-carbons leading to a calculated shift

$$\delta_{C3} = 123.3 + 10.6 + 7.2 - 7.9 - 1.8 = 131.4$$

which would mean that the line at 130.3 in the spectrum is certainly due to the 3-carbon. In evaluating these shifts from this expression, further vinyl carbons are treated just as are alkyl carbons, since conjugation has little effect on alkene chemical shifts. The effect of introducing other substituents is shown in Table 6.3.

While there is little change in chemical shift between conjugated and unconjugated dienes or indeed even between alkenes and aromatic systems, allenes show a most striking chemical shift for the central sp carbon, as low as 208 for C-2 in the case of 1,2-pentadiene.

Table 6.3 Effect of Various Functional Groups on Alkene Chemical Shifts[a]

Group	β	α	α'	β'
OR	2	29	−39	−1
OH	6	—	—	−1
OAc		18	−27	
COCH₃		15	6	
CHO		13	13	
COOH		4	9	
COOR		6	7	
CN		−16	15	
Cl	−1	3	−6	2
Br	0	−8	−1	2
I		−38	7	
C₆H₅		12	−11	

Steric Correction Terms for Pairs of Substituents

α, α'	*trans*	0
α, α'	*cis*	−1.1
α, α		−4.8
α', α'		+2.5
β, β		+2.3

[a] To be added to basic alkene shift of δ123.3.

ALKYNES

Alkynes have chemical shifts midway between alkanes and alkenes, in the region of 75–95 ppm. The carbons are often readily assigned, as is the case in the spectrum of 1-hexyne in Figure 6.11. Clearly the line at δ84.5 is C-2. Note that we expect it to be less intense; since it has no attached hydrogens it will exhibit much less NOE.

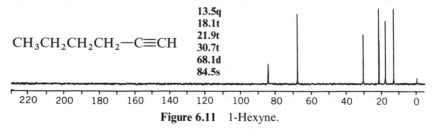

$CH_3CH_2CH_2CH_2—C\equiv CH$

13.5q
18.1t
21.9t
30.7t
68.1d
84.5s

Figure 6.11 1-Hexyne.

AROMATIC CARBONS

Aromatic carbons are not influenced by the "ring current" that deshields the aromatic protons. Their chemical shifts are thus in the same ranges as alkenes, 110–133 ppm. They can often be distinguished from alkene carbons by their longer T_1s whose measurement is discussed later in the chapter.

The chemical shift of benzene carbons is $\delta128.5$, and their shifts are linearly influenced by substituents. Some of the substituent constants are shown in Table 6.4.[5] The carbons bearing the substituent can often be recognized by their substantially lower intensity and by their remaining singlets upon off-resonance decoupling.

Let us consider the spectrum of *ortho*-bromoanisole in Figure 6.12. The observed shifts are $\delta56.0$, 111.7, 112.2, 121.8, 128.5, 133.2, and 156.0. The

Table 6.4 Values to Add^a in Assigning Chemical Shifts in Substituted Benzenes[5]

Group	C-1	ortho	meta	para
H	0	0	0	0
CH_3	9.3	0.8	0	−2.9
CH_2CH_3	15.6	−0.4	0	−2.6
$CH(CH_3)_2$	20.2	−2.5	0.1	−2.4
$C(CH_3)_2$	22.4	−3.1	−0.1	−2.9
C_6H_5	13	−1	0.4	−1
$CH{=}CH_2$	9.5	−2.0	0.2	−0.5
$C{\equiv}CH$	−6.1	3.8	0.4	−0.2
CH_2OH	12	−1	0	−1
COOH	2.1	1.5	0	5.1
$COOCH_3$	2.1	1.1	0.1	4.5
CHO	8.6	1.3	0.6	5.5
$COCH_3$	9.1	0.1	0	4.2
CN	−15.4	3.6	0.6	3.9
OH	26.9	−12.7	1.4	−7.3
OCH_3	31.4	−14.4	1.0	−7.7
NH_2	18.0	−13.3	0.9	−9.8
$N(CH_3)_2$	23	−16	1	−12
NO_2	20.0	−4.8	0.9	5.8
F	34.8	−12.9	1.4	−4.5
Cl	6.2	0.4	1.3	−1.9
Br	−5.5	3.4	1.7	−1.6
I	−32	10	+3	+1

^{a} To be added to the chemical shift of benzene, $\delta128.5$.

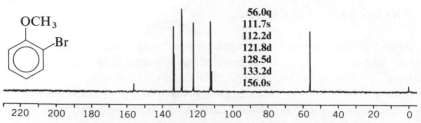

Figure 6.12   *o*-Bromoanisole, (2-bromophenyl methyl ether).

line at δ56.0 can clearly be assigned to the CH_3, and the two of lesser intensity at δ156.0 and δ111.7 must be due to the C—Br and C—OCH_3, although not necessarily in that order.

The remaining calculated lines and their calculated assignments are:

	Experimental	Calculated
C1 (OCH_3)	156.0	156.5
C2 (Br)	111.7	108.6
C3	133.2	132.8
C4	121.8	120.4
C5	128.5	127.9
C6	112.2	115.8
C7 (CH_3)	56.0	—

These additive parameters work very well as assignment guides unless there are two *ortho* substituents that interact. The effect of phenyl groups on alkyl chains is shown in Table 6.6 and on alkenes in Table 6.3.

CHARGED AROMATIC SYSTEMS

The variation in chemical shifts with charge that was observed in pmr is also found in ¹³C spectra. In both cases an upfield shift is observed with increased π-electron density:

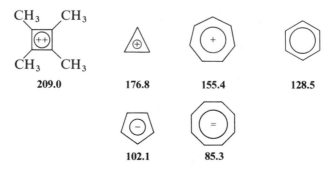

Analysis of these data led Spiesecke and Schneider[6] to predict the constant of 160 ppm/charge for ^{13}C cyclic π-systems. This compares with the smaller pmr shift of 10 ppm/charge.

CARBONYL CARBONS

Carbonyl carbons are strongly deshielded, even compared to olefinic carbons.[7]

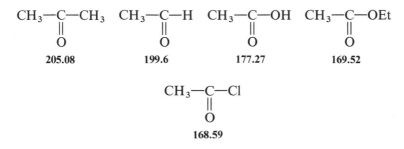

205.08	199.6	177.27	169.52

$$CH_3-\underset{\underset{O}{\|}}{C}-Cl$$

168.59

This is presumably because of the C—O bond polarization:

Any functional group that introduces more electron density on the carbonyl carbon thus will tend to cause an upfield shift, as is evidenced by comparing the shifts in acetone and acetic acid above.

A more complete list of variations in carbonyl chemical shift with substituent is given in Table 6.5. Since the chemical shift depends on the efficiency of conjugation with electrons and π-systems on the substituent, co-planarity of π-systems is required. This is shown below where introduction of methyl groups on the aromatic ring prevents co-planarity.[8]

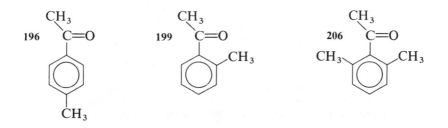

Table 6.5   Variation in Carbonyl Shift
with Substituent Group X:

$$CH_3-\underset{\underset{O}{\|}}{C}-X$$

X	δ_{CO}
H	199.6
CH_3	205.08
Ph	196.0
$CH=CH_2$	197.2
OH	177.27
OCH_3	170.7
NMe_2	169.63
Cl	168.59
Br	165.65
I	158.9

Carbonyls affect the chemical shifts of adjacent aryl carbons and alkenyl carbons as shown in Tables 6.4 and 6.3. Their effect on alkyl carbons is shown in Table 6.6. Their extreme deshielding makes them easy to recognize, but they may occasionally not be observed in cases of extremely rapid pulsing because of their long T_1's and lack of NOE. This low intensity is shown in Figure 6.8 and in Figure 6.13, the spectrum of 3-methyl-2-butanone.

Table 6.6   Effect of Carbonyl and Phenyl Groups on Alkyl Chemical
Shifts[a][5]

Straight: $R-\underset{\alpha}{CH_2}-\underset{\beta}{CH_2}-\underset{\gamma}{CH_2}$ Branched: $\underset{\gamma}{C}-\underset{\beta}{C}-\overset{\overset{\displaystyle R}{|}}{\underset{\alpha}{C}}-\underset{\beta}{C}-\underset{\delta}{C}$

	α		β		γ
R	st	br	st	br	
COOH	21	16	3	2	−2
COOR	20	17	3	2	−2
COCl	33	28	—	2	—
CHO	31	—	0	—	−2
COR	30	24	1	1	−2
Ph	23	17	9	7	−2

[a] To be added to the normal alkyl shift for that carbon.

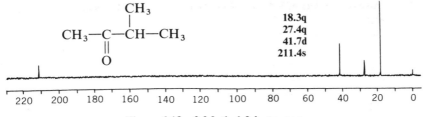

$$18.3q$$
$$27.4q$$
$$41.7d$$
$$211.4s$$

Figure 6.13 3-Methyl-2-butanone.

The use of Table 6.6 is illustrated below for 3-methyl-2-butanone. The table predicts the change in chemical shift in the α-, β-, and γ-carbons when R is substituted for H. For 3-methyl-2-butanone, the COR group is substituted on methane and propane:

We thus obtain the chemical shifts for methane (-2.1) and C-1 and C-2 of propane (15.6 and 16.1) from the beginning of the chapter and calculate the modification of the shift when COR is substituted.

	Calculated	Experimental
(using straight chain) $\delta_1 = -2.1 + 30 = 27.9$		27.4
(using branched) $\delta_3 = 16.6 + 24 = 40.1$		41.7
(using branched) $\delta_4 = 15.6 + 1 = 16.6$		18.3

The lines in the experimental spectrum are found at $\delta 18.3, 27.4, 41.7,$ and 211.4. We can thus assign $\delta 18.3$ to C4, $\delta 41.7$ to C3, $\delta 27.4$ to C1, and by elimination $\delta 211.4$ to C2.

ALCOHOLS, ETHERS, AND HALOGENS

Alcohol, ether, and halogen groups affect the chemical shifts of α-, β-, and γ-carbons as shown in Table 6.7. Note as before that these are the *deviations* from the chemical shifts calculated (or observed) for the parent alkanes. In

Table 6.7 Effect of X-Group Substitution on Alkyl Chemical Shifts[5]

X	α		β		γ
	st	br	st	br	
OH	48	41	10	8	−5
OR	58	51	8	5	−4
OC—R $\overset{\|}{\underset{O}{}}$	51	45	6	5	−3
F	68	63	9	6	−4
Cl	31	32	11	10	−4
Br	20	25	11	10	−3
I	−6	4	11	12	−1

particular, note that an iodine substituent causes an *up field* shift of the carbon it is attached to. Here the back-donation of electrons from the iodine atom is more important than its electronegativity.

NITROGEN AND SULFUR GROUPS

Table 6.8 shows the changes in chemical shift from the parent alkane for nitro, amino, cyano, and thio groups.

Table 6.8 Effect of Nitrogen and Sulfur Groups on Alkyl Chemical Shifts[5]

R	α		β		γ
	st	br	st	br	
NH_2	29	24	11	10	−5
NH_3	26	24	8	6	−5
NHR	37	31	8	6	−4
NR_2	42	—	6	—	−3
NO_2	63	57	4	4	—
CN	4	1	3	3	−3
SH	11	11	12	11	−4
SR	20	—	7	—	−3

MEASUREMENT OF SPIN–LATTICE RELAXATION TIMES

The spin–lattice relaxation time, T_1, is a characteristic part of each line in an nmr spectrum and is particularly instructive in ^{13}C spectroscopy, where the protons are usually decoupled from the carbons. Since the spin–lattice relaxation time is related to how often pulses can be applied to the sample, it is possible to measure the various T_1s in a spectrum all at once using a simple two-pulse sequence.[9]

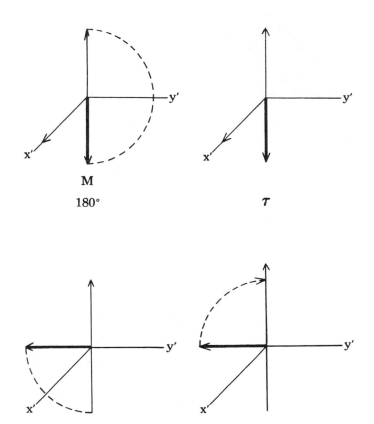

$\tau \ll T_1$

Figure 6.14 Inversion-recovery T_1 experiment when $\tau \ll T_1$.

While there are a number of methods for measuring T_1s now in use, the simplest and most popular is the inversion–recovery sequence described by

$$-(-180-\tau-90-T-)-_n$$

where τ is a variable delay time, T is a long, fixed delay greater than five times the longest T_1 to be measured, and the 180 and 90 are two rf pulses. To understand how this sequence works, let us consider first the case where the time between the two pulses is very short compared to T_1. The 180° pulse will completely invert the magnetization to the $-z$ axis as shown in Figure 6.14 and the 90° pulse which quickly follows will cause the magnetization to rotate further to the $-y$ axis. Since together these are the equivalent of a 270° pulse, the signal that is acquired after the 90° pulse will be exactly the negative of that observed after a 90° pulse by itself. Thus the transformed spectrum would have all of its lines inverted.

Now let us consider a second case in which the delay time is somewhat longer. After the 180° pulse, the nuclei have some time to begin relaxation before the second pulse is applied as shown in Figure 6.15. They relax according to the spin–lattice mechanism up through the z-axis, and the magnetization along the $-z$-axis becomes less intense. When the 90° pulse

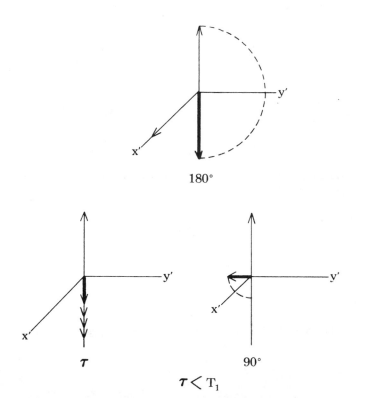

Figure 6.15 Inversion-recovery experiment when $\tau < T_1$.

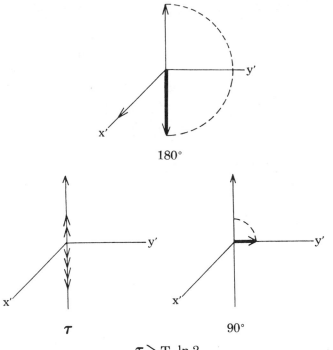

$$\tau > T_1 \ln 2$$

Figure 6.16 Inversion-recovery experiment when $\tau > T_1 \ln 2$.

is applied after this longer delay, a much smaller amount of magnetization is rotated into the x–y plane, leading to a smaller, but still negative, signal.

In a third case where τ is longer still and is greater than $T_1 \ln 2$, the magnetization will have to relax further between the pulses and will be pointing in the positive z-direction before the 90° pulse is applied. The result will be a small, positive signal. This is shown in Figure 6.16.

Finally, if the time between the 180° and 90° pulse is much greater than T_1, there will be plenty of time for the magnetization to regain its equilibrium value before the second pulse is applied, and a fully positive fid will be observed, just as if the 180° pulse had never been applied.

The reason that this is such a powerful technique lies in the fact that just as in single pulse experiments, the T_1 relaxation times of all the lines in the spectrum can be observed in this way. Such spectra for a series of inversion–recovery experiments with differing τ values are shown for dodecanol in Figure 6.17 as a stacked plot of a series of experiments.

More important than these impressive plots, however, is the fact that the minicomputer can actually calculate the T_1s of each line in the spectrum from the intensities or areas and the list of delay times τ.[10] The intensities of the lines in an inversion–recovery experiment are given by

$$A_i = A_\infty \left[1 - 2 \exp\left(\frac{-\tau_i}{T_1}\right) \right] \tag{6.4}$$

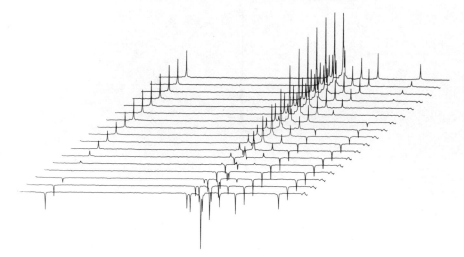

Figure 6.17 Typical data from an inversion-recovery T_1 experiment. The spectrum is of dodecyl alcohol in deuterobenzene with delay times of 0.1, 0.2, 0.3, 0.4, 0.5, 0.6, 0.7, 0.8, 0.9, 1.0, 1.5, 2.0, 2.5, 3.0, 3.5, 4.0, 6.0, 8.0, and 10.0 sec.

where A_i is the intensity of the line in the ith spectrum, τ_i is the delay time used in acquiring that spectrum, and A_∞ is the intensity of the line in the "infinity" spectrum where τ is much greater than T_1. It has been recognized that simple logarithmic plots of these data cast in the form

$$\ln\left(\frac{1-A}{A_\infty}\right) = \frac{-\tau_i}{T_1} + \ln 2 \tag{6.5}$$

weight the τ_∞ spectrum excessively, and exponential fits are more commonly used.[11]

REASONS FOR MEASURING T_1s

Organic chemists have become interested in the measurement of T_1s primarily because they can impart additional structural information in many cases. Lines having different T_1s but similar chemical shifts can sometimes be assigned by the measurement of their T_1s. Similarly, Allerhand has shown that T_1s can be used to assign the chemical shifts of lines belonging to similar functional groups where one is buried within the molecule and another lies near the outside, even when these fragments are monosaccharides.[12] The reason that such assignments can be made is that the value of T_1 for a given

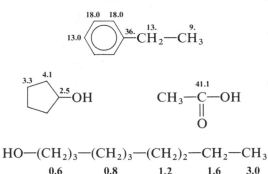

Figure 6.18   Spin–lattice relaxation times (T_1s) for common compounds.

spectral line is inversely related to the *rotational correlation time* for that functional group. This is illustrated below for nitrobenzene:[13]

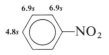

The nitrobenzene molecule presumably tumbles with substantial rotation about the NO$_2$—C$_4$ axis. Thus the C$_2$ and C$_3$ carbons are rotating faster and tumbling more freely than the C$_4$ carbon. They therefore have *longer* T_1s than C$_4$ since T_1 is inversely related to this tumbling rate.

Relaxation measurements have also been used to investigate segmental motion in long chain molecules that may serve as biological models. For example, the C-10 carbon of 1-decanol[14] (Figure 6.18) has been shown to have a much longer T_1 than C-1, because it can move more freely, while the C-1 carbon is more constrained by hydrogen bonding.

In general, aliphatic carbons have rather short T_1s and aromatic and carbonyl carbons have longer T_1s. In addition, chlorinated carbons such as CDCl$_3$ have longer T_1s as illustrated in Figure 6.8. Carbons bound to nuclei other than ^{1}H will have longer T_1s (Br is the only exception). A few typical T_1s are shown for various molecules in Figure 6.18. A complete review of uses of T_1s as structural tools has been given by Wehrli.[15]

PROBLEMS IN ^{13}C NMR SPECTROSCOPY

The following problems are a series of ^{13}C spectra with the chemical shifts listed on them. Most are plotted on the same 230-ppm scale as are the ones in the chapter, but look at the scale to be sure. The chemical shifts are listed followed by the multiplicity of the line when an off-resonance decoupled spectrum is obtained (s = singlet, d = doublet, t = triplet, etc.). In a few cases, the intensities of the various lines are also given so that all of the peaks can be identified when they are very close together. You should try (a) to identify the compound and (b) to assign all the lines in the spectrum to justify your assignment. Use the tables and equations in the chapter to prove that your assignments are correct.

PROBLEMS

Problem 6.1 C_8H_{18}

24.9d
25.5q
30.2q
31.2s
53.4t

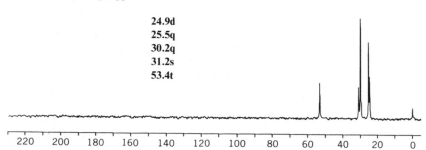

Problem 6.2 $C_5H_{11}Cl$

22.0q
25.7d
41.6t
43.1t

Problem 6.3 $C_4H_8O_2$

14.4q
20.9q
60.4t
170.7s

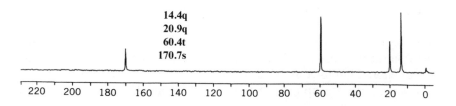

Problem 6.4 C_4H_7Br

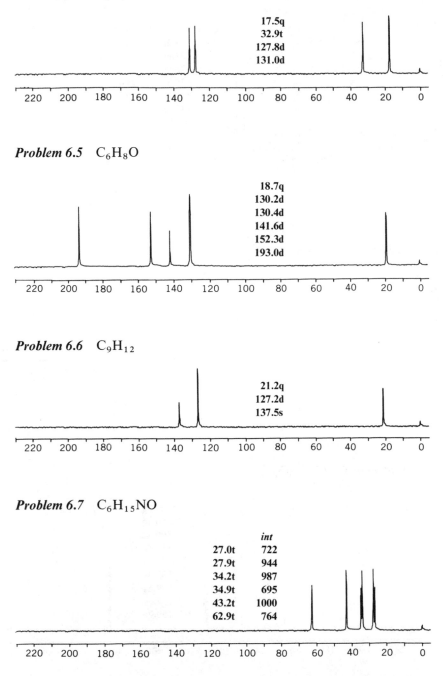

17.5q
32.9t
127.8d
131.0d

Problem 6.5 C_6H_8O

18.7q
130.2d
130.4d
141.6d
152.3d
193.0d

Problem 6.6 C_9H_{12}

21.2q
127.2d
137.5s

Problem 6.7 $C_6H_{15}NO$

	int
27.0t	722
27.9t	944
34.2t	987
34.9t	695
43.2t	1000
62.9t	764

Problem 6.8 $C_9H_{10}O$

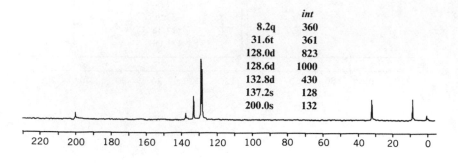

	int
8.2q	360
31.6t	361
128.0d	823
128.6d	1000
132.8d	430
137.2s	128
200.0s	132

Problem 6.9 $C_5H_8O_2$

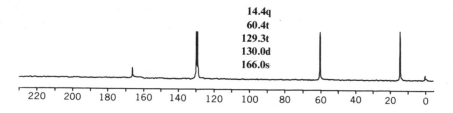

14.4q
60.4t
129.3t
130.0d
166.0s

Problem 6.10 $C_9H_{10}O_3$

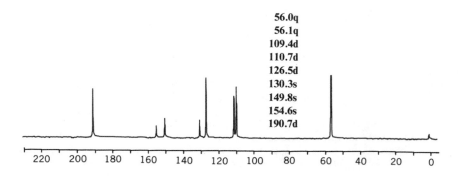

56.0q
56.1q
109.4d
110.7d
126.5d
130.3s
149.8s
154.6s
190.7d

Problem 6.11 $C_{12}H_{14}O_4$

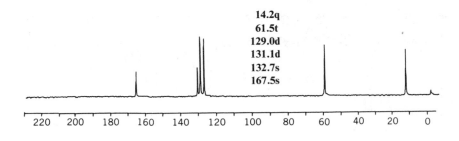

14.2q
61.5t
129.0d
131.1d
132.7s
167.5s

Problem 6.12 $C_9H_8O_3$

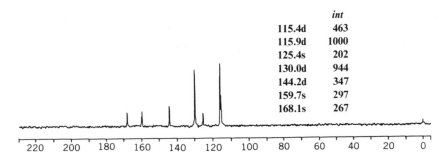

	int
115.4d	463
115.9d	1000
125.4s	202
130.0d	944
144.2d	347
159.7s	297
168.1s	267

Problem 6.13 $C_8H_{11}N$

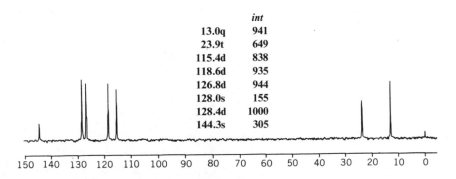

	int
13.0q	941
23.9t	649
115.4d	838
118.6d	935
126.8d	944
128.0s	155
128.4d	1000
144.3s	305

Problem 6.14 ¹H nmr, IR and ¹³C nmr spectra of compound with formula C_7H_8

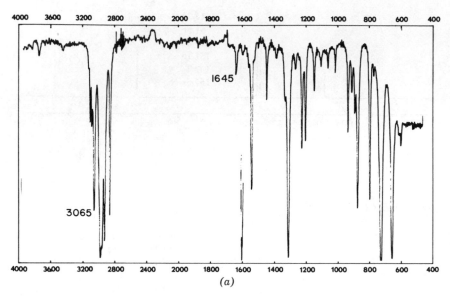

(a)

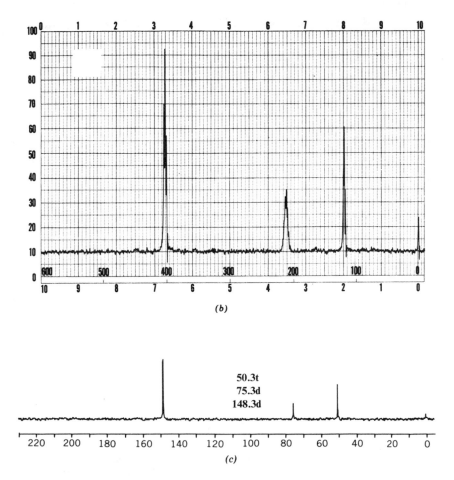

(b)

50.3t
75.3d
148.3d

(c)

Problem 6.15 ¹H nmr, IR and ¹³C nmr spectra of compound with formula $C_5H_{11}N$

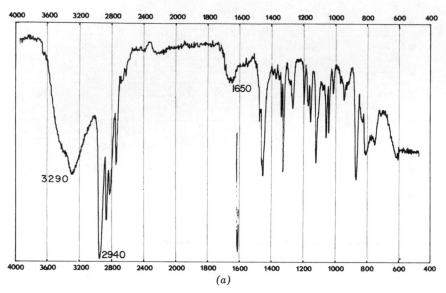

(a)

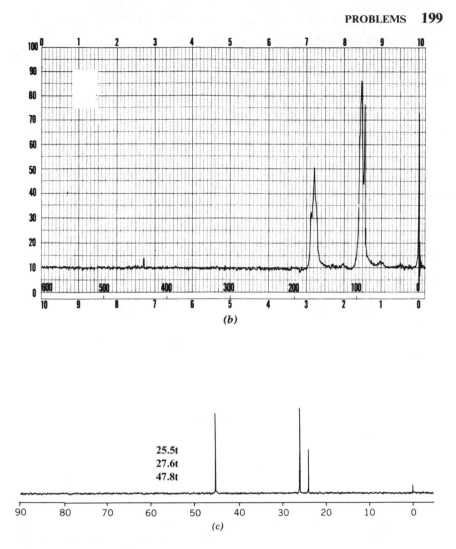

(b)

25.5t
27.6t
47.8t

(c)

RECOMMENDED READING

Abraham, R. J., and P. Loftus, *Proton and Carbon-13 Nmr Spectroscopy: An Integrated Approach*, Heyden and Son, New York, 1973.

Levy, G. C., and G. L. Nelson, *Carbon-13 Nuclear Magnetic Resonance for Organic Chemists*, Wiley-Interscience, New York, 1972.

Müllen, K., and P. S. Pregosin, *Fourier Transform Nmr Techniques: A Practical Approach*, Academic Press, New York, 1976.

Shaw, D., *Fourier Transform Nmr Spectroscopy*, Elsevier, New York, 1976.

Stothers, J. B., *Carbon-13 Nmr Spectroscopy*, Academic Press, New York, 1972.

Wehrli, F. W., and T. Wirthlin, *Interpretation of Carbon-13 Nmr Spectra*, Heyden and Son, New York, 1976.

SPECTRAL COLLECTIONS

Breitmaier, E., Hass, G., and Voelter, W., *Atlas of Carbon-13 Nmr Data*, Vol. 1, IFI/Plenum, New York, 1975.

Formacek, V., L. Desnoyer, H. P. Kellerhals, T. Keller, and J. T. Clerc, ^{13}*C Data Bank*, Vol. 1, Bruker-Physik, Karlsruhe, 1976.

Johnson, L. F., and W. C. Jankowski, *Carbon-13 Nmr Spectra*, Wiley-Interscience, New York, 1972.

REFERENCES

1. J. H. Noggle and R. E. Schirmer, *The Nuclear Overhauser Effect—Chemical Applications*, Academic Press, New York, 1971.
2. D. M. Grant and E. G. Paul, *J. Am. Chem. Soc.*, **86**, 2984 (1964).
3. L. P. Lindeman and J. Q. Adams, *Anal. Chem.*, **43**, 1245 (1971).
4. D. E. Dorman, M. Jautelat, and J. D. Roberts, *J. Org. Chem.*, **36**, 2757 (1971).
5. F. W. Wehrli and T. Wirthlin, *Interpretation of Carbon-13 Nmr Spectra*, Heyden and Son, New York, 1976, pp. 47 ff.
6. H. Spiesecke and W. G. Schneider, *Tetrahedron Lett.*, 468 (1961).
7. J. B. Stothers, *Carbon-13 Nmr Spectra*, Academic Press, New York, 1972, p. 280.
8. F. W. Wehrli and T. Wirthlin, *Interpretation of Carbon-13 Nmr Spectra*, Heyden and Son, New York, 1976, p. 34.
9. R. L. Vold et al., *J. Chem. Phys.*, **48**, 3831 (1968).
10. J. W. Cooper, Amer. Laboratory, **5** (9), 63 (Sept. 1973).
11. J. Kowalewski et al., *J. Magn. Resonance*, **26**, 533 (1977).
12. A. Allerhand and D. Doddrell, *J. Am. Chem. Soc.*, **93**, 277 (1971).
13. G. C. Levy, J. D. Cargoli, and F. A. L. Anet, *J. Am. Chem. Soc.*, **95**, 1527 (1973).
14. D. Doddrell and A. Allerhand, *J. Am. Chem. Soc.*, **93**, 1558 (1971).
15. F. Wehrli, " Organic Structure Assignments Using C-13 Spin-Relaxation Data," in *Topics in Carbon-13 Nmr Spectroscopy*, Vol. 2, G. Levy, Ed., Wiley-Interscience, New York, 1976, p. 343.

CHAPTER SEVEN

SIMPLE HÜCKEL MO THEORY

CLASSICAL DESCRIPTION OF WAVES

In this chapter we will present a very simple technique for describing the energies and resonance stabilization of conjugated hydrocarbons. This technique is called Simple Hückel Molecular Orbital (SHMO) Theory and can be used to predict the stability of conjugated systems and to correlate it with various spectral properties. We will also find that it can predict reactivities and bond strengths. This theory treats electrons as waves moving in a conjugated π-system. To understand the origins of this theory, due to Ernst Hückel (1922), we will start by considering waves in classical physics.

Let us consider a cork bobbing on a series of waves in the water. Its position g will vary as a function of time t and as a function of the distance between waves x. We can write this mathematically by saying

$$g(x, t) = \text{some expression.}$$

When we talk about functions describing waves, it has been conventional to use the Greek symbol ψ (psi), and we then write

$$\psi(x, t) = \text{some expression.}$$

We will now develop an expression describing that cork from elementary physics. By Hooke's law, we can describe the force necessary to restore that cork to zero amplitude by

$$f = -k\psi. \tag{7.1}$$

Since

$$f = ma \tag{7.2}$$

and acceleration is just the second derivative of the position ψ with respect to time, we can write

$$f = m\frac{\partial^2 \psi}{\partial t^2} = -k\psi, \tag{7.3}$$

201

This, of course, is a differential equation. One solution is

$$\psi = A \sin\left[\left(\frac{k}{m}\right)^{1/2} t\right] \tag{7.4}$$

where A is a constant.

We will now eliminate time from this expression to arrive at a *stationary state* description of the function. The units of $(k/m)^{1/2}$ are sec^{-1} and represent the angular velocity of the wave. We can substitute $2\pi v$ for this expression giving

$$\psi = A \sin(2\pi vt). \tag{7.5}$$

In describing the cork's position, we might find it convenient to refer to it with respect to some arbitrary time τ, and write our expression as

$$\psi = A \sin[2\pi v(t - \tau)]. \tag{7.6}$$

The velocity of any wave c is simply the distance x traveled per unit time t

$$c = \frac{x}{t} = v\lambda \tag{7.7}$$

and thus

$$\frac{x}{\lambda} = vt. \tag{7.8}$$

We can replace $v\tau$ with x/λ in Eq. 7.6, yielding

$$\psi = A \sin 2\pi\left(\frac{vt - x}{\lambda}\right). \tag{7.9}$$

To get rid of the time dependence in this expression, we will perform the common trick of taking the second derivative of the function and discover that the sine portion looks much the same as before:

$$\frac{\partial \psi}{\partial x} = -\frac{2\pi}{\lambda} A \cos 2\pi\left(\frac{vt - x}{\lambda}\right) \tag{7.10}$$

and

$$\frac{\partial^2 \psi}{\partial x^2} = \frac{-4\pi^2}{\lambda^2} A \sin 2\pi\left(vt - \frac{x}{\lambda}\right). \tag{7.11}$$

Inspecting Eqs. 7.9 and 7.11, we find that

$$\frac{\partial^2 \psi}{\partial x^2} = \frac{-4\pi^2}{\lambda^2} \psi. \tag{7.12}$$

We have now eliminated t and arrived at a stationary state description of the *wave function* affecting our cork.

In a similar manner, in three dimensions, we could write a wave function

$$\frac{\partial^2 \psi}{\partial x^2} + \frac{\partial^2 \psi}{\partial y^2} + \frac{\partial^2 \psi}{\partial z^2} = \frac{-4\pi^2}{\lambda^2} \psi. \tag{7.13}$$

So that we do not have to write out all those partial derivatives every time, we symbolize the three second partials by the symbol ∇^2, called "del-squared." This simplifies Eq. 7.13 to

$$\nabla^2 \psi = \frac{-4\pi^2}{\lambda^2} \psi. \tag{7.14}$$

QUANTUM MECHANICAL CONSIDERATIONS

We will now begin to talk about rather smaller particles than corks and find that we can express their behavior in a similar fashion. In 1924, Louis de Broglie postulated that if light has a particle nature as well as a wave nature, then we might expect matter to have both a particle nature and a wave nature as well. Thus he reasoned that any mass moving at some velocity must have a wavelength associated with it. Since he knew that

$$E = h\nu = mc^2 \qquad \text{and} \qquad \nu = \frac{c}{\lambda}$$

he reasoned that

$$\lambda = \frac{h}{mc}. \tag{7.15}$$

Thus any moving particle has an associated *de Broglie wavelength*. Some typical wavelengths are shown below:

Particle	m	c (cm/sec)	λ (cm)
Slow electron	e	1	7.27
Fast electron	e	6×10^8	10^{-8} (1Å)
Tennis ball	42 g	2500	5.7×10^{-32}

Rewriting Eq. 7.14 using the fact that

$$\frac{1}{\lambda^2} = \frac{m^2 c^2}{h^2} \tag{7.16}$$

we have

$$\nabla^2 \psi = \frac{-4\pi^2 m^2 c^2}{h^2} \psi. \tag{7.17}$$

Since the total energy in any system is the sum of the kinetic energy T and the potential energy V,

$$E_{\text{tot}} = \text{kinetic} + \text{potential} = T + V$$

then

$$T = E - V = \frac{mc^2}{2} \tag{7.18}$$

or

$$c^2 = \frac{2}{m}(E - V). \tag{7.19}$$

We can now eliminate the velocity c from Eq. 7.17 in favor of energy terms

$$\nabla^2 \psi = \frac{-4\pi^2 m^2}{h^2}\left(\frac{2}{m}\right)(E - V)\psi$$

or

$$\nabla^2 \psi = \frac{-8\pi^2 m}{h^2}(E - V)\psi. \tag{7.20}$$

This is the *Schroedinger wave equation*. If we rearrange it to the following form

$$-\frac{h^2}{8\pi^2 m}(\nabla^2 + V)\psi = E\psi \tag{7.21}$$

we can designate all the constants and the operator ∇^2 by a new symbol $\mathscr{H}$ and write the simple final equation

$$\mathscr{H}\psi = E\psi. \tag{7.22}$$

We see that $\mathscr{H}$ specifies a set of mathematical operations to be done on the wave function ψ and contains both kinetic and potential energy terms. The wave function ψ, on the other hand, is now an *electron* amplitude function, which might be positive or negative at any point in space.

The value of $\psi^2\, dx\, dy\, dz$ is the probability of finding an electron in the small space $dx\, dy\, dz$ and therefore it is essential that

$$\iiint_{-\infty}^{+\infty} \psi^2\, dx\, dy\, dz = 1 \qquad (7.23)$$

since the probability of finding the electron in all space must be exactly unity. We can write this as $\int \psi^2\, d\tau = 1$, where we imply a triple integral over all space by $d\tau$.

SOLUTIONS OF THE SCHROEDINGER WAVE EQUATION

The Schroedinger wave equation describes the motion of a single electron only. Thus it actually applies only to the hydrogen atom. There are many possible solutions to the equation, including the trivial one of

$$\psi = 0.$$

However, for useful solutions, we must have single-valued, finite, and continuous expressions. For each such expression

$$\mathscr{H}\psi_i = E_i\,\psi_i$$

where ψ_i, a particular solution, is called an *eigenfunction* and E_i, the energy for that solution, is called an *eigenvalue*. Each solution is of the form $\psi(\phi, \theta, r, n, l, m)$, where ϕ, θ, and r are spherical coordinates and n, l, and m are integers called *quantum numbers*. It is these quantum numbers, of course, that lead to spatial descriptions of the various orbitals we use daily in organic chemistry.

The quantum number n corresponds to the *shell* that the electrons inhabit, or to their row in the periodic table. It can vary from 1 to 6 for the known elements. The hydrogen atom contains only $1s$ orbitals and thus has n equal to 1 for its electron. Carbon atoms have both $1s$ and $2s$ electrons and these outer bonding electrons ($2s$) have $n = 2$.

The quantum number l can vary from 0 to $n - 1$ and corresponds to which of the suborbitals the electron can inhabit. Thus $l = 0$ corresponds to the s level, $l = 1$ corresponds to the p-orbitals, and $l = 2$ and 3 correspond to the d- and f-orbitals.

Finally the m quantum number is used to select which of the several possible suborbitals the electron belongs in and can vary from $-l$ to $+l$. The p_x, p_y, and p_z-orbitals of the second row correspond to m quantum numbers of -1, 0, and $+1$. Thus $n = 2$, $l = 1$, and $m = 0$ defines the wave function for the $2p_y$-orbital.

For the hydrogen atom, we know that the energies associated with these quantum numbers go up with increasing n, but for multielectron atoms, the electron interaction complicates the potential energy term with repulsion interactions.

SIMPLE MOLECULAR ORBITAL THEORY

For our purposes, however, we will ignore these potential interactions and, in fact, will ignore all s-orbitals, and will concern ourselves with electron waves only in π-systems. We will treat each electron as circulating by itself, seeing only an effective nuclear charge of $+1$.

As a first simple problem, we will consider H_2^+, the hydrogen molecule ion. It consists of one electron but two atoms. We would like to describe a single molecular orbital (MO) with one electron in it. The principle assumption here and throughout this treatment will be that *we can describe such a molecular orbital as a linear combination of atomic orbitals*. We call this an LCAO–MO approach.

CALCULATION OF THE MO FOR H_2^+

We start by writing down our wave equation

$$\mathscr{H}\psi = E\psi. \tag{7.22}$$

Multiplying, we have

$$\psi\mathscr{H}\psi = E\psi^2. \tag{7.24}$$

Note that $\psi\mathscr{H}\psi \neq \mathscr{H}\psi^2$ since $\mathscr{H}$ *operates* on the function to its right, but is *multiplied* by any function to its left. Integrating, we have

$$\int \psi\mathscr{H}\psi \, d\tau = E \int \psi^2 d\tau.$$

The wave function we are trying to solve will be a linear combination of atomic wave functions, each multiplied by a constant c_i:

$$\psi = \sum_{i=1}^{n} c_i\psi_i. \tag{7.26}$$

We want to solve for these multiplying constants to determine what fraction of each atomic wave function we will mix together with the others to define the molecular wave function for the molecule we wish to describe.

So, substituting Eq. 7.26 in 7.25, we have

$$\int \sum c_i \psi_i \mathscr{H} \sum c_i \psi_i \, d\tau - E \int [\sum c_i \psi_i]^2 \, d\tau = 0. \qquad (7.27)$$

Rewriting this, we have

$$\sum_{i=1}^{n} \sum_{j=1}^{n} c_i c_j \left[\int \psi_i \mathscr{H} \psi_j \, d\tau - E \int \psi_i \psi_j \, d\tau \right] = 0 \qquad (7.28)$$

where n = the number of atomic orbitals we are using. We will abbreviate even further to simplify our notation, letting

$$\int \psi_i \mathscr{H} \psi_j = H_{ij} \qquad (7.29)$$

and

$$\int \psi_i \psi_j = S_{ij}. \qquad (7.30)$$

Furthermore, for all meaningful solutions we can assume that

$$H_{ij} = H_{ji} \qquad (7.31)$$

$$S_{ij} = S_{ji} \qquad (7.32)$$

So Eq. 7.28 simplifies further to

$$\sum_{i=1}^{n} \sum_{j=1}^{n} c_i c_j (H_{ij} - ES_{ij}) = 0. \qquad (7.33)$$

The crux of the solution is given by the Ritz Variational theorem which we will state without proof:

Any wave function ψ_{wrong} must have an associated energy E_{wrong} which is greater than E_{actual}.

Thus by minimizing the energy, we can solve for the correct ψ. To do this, we set

$$\frac{\partial E}{\partial c_i} = 0$$

for each c_i giving us n equations in n unknowns, c_i.

For the hydrogen molecule ion H_2^+, we want to make up our MO out of the two hydrogen atom orbital functions ψ_1 and ψ_2. Thus we can expand the summation in Eq. 7.33 to

$$c_1 c_2 (H_{11} - ES_{11}) + c_1 c_2 (H_{12} - ES_{12}) + c_2 c_1 (H_{21} - ES_{21})$$
$$+ c_2 c_2 (H_{22} - ES_{22}) = 0. \qquad (7.34)$$

To describe the wave function for H_2^+ we must evaluate the constants c_1 and c_2. We take the partial by c_1 of the Eq. 7.34:

$$\frac{\partial}{\partial c_1}[c_1^2(H_{11} - ES_{11}) + 2c_1c_2(H_{12} - ES_{12}) + c_2^2(H_{22} - ES_{22})] = 0. \quad (7.35)$$

Note that E is the only variable with respect to c_1 and all other numbers are constants. Then

$$2c_1(H_{11} - ES_{11}) + 2c_2(H_{12} - ES_{12}) = 0 \quad (7.36)$$

and similarly for $\partial/\partial c_2$ we get

$$2c_1(H_{12} - ES_{12}) + 2c_2(H_{22} - ES_{22}) = 0. \quad (7.37)$$

This gives two equations in two unknowns c_1 and c_2, which we can write as the determinant

$$\begin{vmatrix} H_{11} - ES_{11} & H_{12} - ES_{12} \\ H_{21} - ES_{21} & H_{22} - ES_{22} \end{vmatrix} = 0. \quad (7.38)$$

These are called *secular equations* and the determinant is called a *secular determinant*.

Once we know E, we can find the ratios of c_1 and c_2. For a general MO we will always get such a secular determinant, of the form

$$\begin{vmatrix} H_{11} - ES_{11} \cdots\cdots H_{1n} - ES_{1n} \\ H_{n1} - ES_{n1} \cdots\cdots H_{nn} - ES_{nn} \end{vmatrix} = 0. \quad (7.39)$$

These always have a diagonal of symmetry and have n real roots. To solve these, we must evaluate H_{ij} and S_{ij}.

S_{ij}—THE OVERLAP INTEGRAL

Recall that we have used S_{ij} to abbreviate the expression $\int \psi_i \psi_j d\tau$ (Eq. 7.30). There are two cases to evaluate. If

$$i = j$$

then

$$S_{ij} = \int \psi_i \psi_i d\tau = \int \psi^2 \, d\tau = 1.$$

Since this agrees with the condition that the integral over all space must be 1 (Eq. 7.23), we call this our *normalization* condition. This simplifies Eq. 7.38 to

$$\begin{vmatrix} H_{11} - E & H_{12} - S_{12}E \\ H_{12} - S_{12}E & H_{22} - E \end{vmatrix} = 0. \quad (7.40)$$

If $i \neq j$, then

$$S_{ij} = \int \psi_i \psi_j \, d\tau.$$

For simplicity, we decide that adjacent wave functions have no interaction or overlap and are therefore *orthogonal*. Mathematically, orthogonal means the wave functions are independent and do not influence one another. We thus say that

$$S_{ij} = 0 \qquad \text{for } i \neq j$$

by our *orthogonality* condition.

Thus the off-diagonal elements become $H_{12} - 0$, and our matrix becomes

$$\begin{vmatrix} H_{11} - E & H_{12} \\ H_{12} & H_{22} - E \end{vmatrix}.$$

THE H INTEGRALS

We have abbreviated as H_{ij}, the quantity $\int \psi_i H \psi_j d\tau$ in Eq. 7.29. Again, there are two cases. If

$$i = j,$$

then

$$H_{ii} = \int \psi_i H \psi_i \, d\tau.$$

We call this the *Coulomb integral*. It is approximately the Coulomb energy of the electron. For further simplification we let

$$H_{ii} = \alpha. \tag{7.41}$$

The value of α is a function of nuclear charge and the type of orbitals involved, but as a 0th approximation, we will assume that α is the same for all hydrogen atoms. As defined here, α is a negative number. If

$$i \neq j$$

we have

$$H_{ij} = \int \psi_i H \psi_j \, d\tau = \beta. \tag{7.42}$$

H_{ij} is the energy of an electron in the fields of atoms i and j involving the wave functions ψ_i and ψ_j. We will call this the *Resonance Integral* and symbolize it by β, with the following strictures:

β has a constant value for all hydrogen atoms at customary bond-forming distances.

$\beta = 0$ if atoms i and j are not nearest neighbors.

FINAL SOLUTION OF H_2^+

The determinant in Eq. 7.40 becomes

$$\begin{vmatrix} H_{11} - S_{11}E & H_{12} - S_{12}E \\ H_{12} - S_{12}E & H_{22} - S_{22}E \end{vmatrix} = \begin{vmatrix} \alpha - E & \beta \\ \beta & \alpha - E \end{vmatrix} = 0. \quad (7.43)$$

This is easily solved by cross multiplying:

$$(\alpha - E)^2 - \beta^2 = 0$$

$$\alpha^2 - 2\alpha E - E^2 - \beta^2 = 0$$

$$E^2 - 2\alpha E + (\alpha^2 - \beta^2) = 0$$

$$E = \frac{2\alpha \pm \sqrt{4\alpha^2 - 4(\alpha^2 - \beta^2)}}{2} = \alpha \pm \beta. \quad (7.44)$$

We now have solved for the energy levels of H_2^+ in terms of a coulombic integral α and a resonance integral β, both of which can be evaluated numerically if desired. Since the measurements of α and β vary with the measurement technique, we often leave the energies in this form. Although it looks like we have solved an equation in several unknowns in terms of two new unknowns, α and β, we *have* progressed in that we now are describing the energy levels of H_2^+ in terms of some *measurable molecular parameters*. Whether we measure them or not we can assume that they are constant for similar molecules, and thus we can compare molecular energies using this MO theory technique.

The equations, in this case, that led to the solution for E were in terms of c_1 and c_2, which we must evaluate in order to express the MO wave function in terms of two H-atom wave functions. These equations were

$$c_1(H_{11} - ES_{11}) + c_2(H_{12} - ES_{12}) = 0 \quad (7.36)$$

$$c_1(H_{12} - ES_{12}) + c_2(H_{22} - ES_{22}) = 0 \quad (7.37)$$

or, substituting Eqs. 7.41 and 7.42 in Eqs. 7.36 and 7.37, we have

$$c_1(\alpha - E) + c_2(\beta - 0) = 0 \tag{7.45}$$

$$c_1(\beta - 0) + c_2(\alpha - E) = 0. \tag{7.46}$$

Rearranging Eq. 7.45, we have

$$\frac{c_1}{c_2} = -\frac{\beta}{\alpha - E} \tag{7.47}$$

and if $E = \alpha + \beta$

$$\frac{c_1}{c_2} = -\frac{\beta}{\alpha - (\alpha + \beta)} = -\frac{\beta}{\alpha - (\alpha + \beta)} = \frac{-\beta}{-\beta} = 1. \tag{7.48}$$

If $E = \alpha - \beta$

$$\frac{c_1}{c_2} = \frac{-\beta}{\beta} = -1. \tag{7.49}$$

We now have the ratio of c_1 and c_2 for $\psi_{\alpha+\beta}$ and $\psi_{\alpha-\beta}$. We could write

$$\psi_I = (1)\psi_1 + (1)\psi_2$$

$$\psi_{II} = (1)\psi_1 + (-1)\psi_2$$

but we must be sure that our new **MO** wave functions are *normalized*. This is the condition that requires that the integral over all space for these wave functions must be 1.

This can be accomplished by setting $c_1 = c_2 = 1/\sqrt{2}$. Then

$$\psi_I = \frac{1}{\sqrt{2}}(\psi_1 + \psi_2) \tag{7.50}$$

$$\psi_{II} = \frac{1}{\sqrt{2}}(\psi_1 - \psi_2) \tag{7.51}$$

Checking to see that ψ_I is normalized, we square it and integrate as we did in Eq. 7.23:

$$\int \psi_I^2 d\tau = \int \left[\frac{1}{\sqrt{2}}(\psi_1 + \psi_2) \right]^2 d\tau = \frac{1}{2}\int \psi_1^2 + \frac{1}{2}\int \psi_2^2 = \frac{1}{2}(1) + \frac{1}{2}(1) = 1.$$

Thus the H$_2^+$ wave function is indeed *normalized*, since the result is 1.

In the H_2^+ molecule, we have two energy levels $\alpha + \beta$ and $\alpha - \beta$. It can be shown that both α and β are negative numbers and thus the lower energy level is $\alpha + \beta$. If we draw the two energy levels and put in the one electron, we have

$$\alpha - \beta \underline{\qquad}$$
$$\alpha + \beta \underline{\quad 1 \quad}.$$

THE HYDROGEN MOLECULE

Our next great leap of faith is this: even when there is more than one electron we ignore electron interactions and consider that this wave functions and energies will be the same. Thus the hydrogen molecule will have exactly the same energies as the hydrogen molecule ion and we fill the lower level with two electrons.

$$\alpha - \beta \underline{\qquad}$$

$$\alpha \quad \text{-------}$$

$$\alpha + \beta \underline{\quad \uparrow\downarrow \quad} \Big\} \text{ resonance stabilization.}$$

The energy of the system is thus $2\alpha + 2\beta$. This is considered to be resonance stabilized by 2β over the hypothetical unbonded energy 2α.

CARBON π-BONDING

We now carry our approximation further by assuming that π-electrons in conjugated hydrocarbons can be described by wave functions similar to those used in the description of the hydrogen molecule. Specifically

1. We will ignore all except π-electrons.
2. We will compare only relative bonding strengths.
3. We will recognize that C—C bonds have a different α and β, but will not in general, evaluate them.
4. We will assume that all other electrons are localized and do not influence the π-bonds.

Despite these approximations, we will discover that our calculated energies will correlate well with measurable quantities.

THE ETHYLENE MOLECULE

Let us consider the ethylene molecule as the simplest carbon π-system. We will construct an MO description of the π-bond in terms of two atomic wave functions.

$$\psi = c_1\psi_1 + c_1\psi_2.$$

Again we will solve the function in the same way as for hydrogen, leading to

$$E_1 = \alpha - \beta \underline{\hspace{2cm}}$$
$$E_2 = \alpha + \beta \underline{\;\uparrow\downarrow\;}$$

where α and β are different from those for H_2, but the same for all carbon π-systems. Consequently, the energy of the ethylene molecule will again be found to be $2\alpha + 2\beta$, where α and β have different values than the H_2.

BUTADIENE

To construct an MO for butadiene, we will need to put together a linear combination of atomic orbitals, one for each atom:

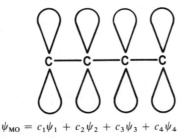

$$\psi_{MO} = c_1\psi_1 + c_2\psi_2 + c_3\psi_3 + c_4\psi_4$$

We can write the same sort of matrix for butadiene as for H_2^+:

$$
\begin{vmatrix}
H_{11} - ES_{11} & \cdots & H_{14} - ES_{14} \\
 & & \\
 & & \\
 & & \\
H_{14} - ES_{14} & \cdots & H_{44} - ES_{44}
\end{vmatrix}
=
\begin{vmatrix}
\alpha - E & \beta & 0 & 0 \\
\beta & \alpha - E & \beta & 0 \\
0 & \beta & \alpha - E & 0 \\
0 & 0 & \beta & \alpha - E
\end{vmatrix}
= 0.
$$

Here the elements of the determinant not referring to adjacent atoms become zero, since $H_{13} = H_{14} = H_{24} = 0$ and all $S_{ij} = 0$ for $i \neq j$. Dividing by β, and letting $\alpha - E/\beta = x$, we have

$$\begin{vmatrix} x & 1 & 0 & 0 \\ 1 & x & 1 & 0 \\ 0 & 1 & x & 1 \\ 0 & 0 & 1 & x \end{vmatrix} = 0.$$

This determinant can be broken down and solved by hand, or it can be solved by a simple computer program, as we will see shortly. By hand, the last steps are:

$$x^4 - 3x^2 + 1 = 0$$

$$x = \pm 1.618$$

$$x = \pm 0.618.$$

Since $x = (\alpha - E)/\beta$, $E = \alpha - \beta x$, and the four energy levels of butadiene are

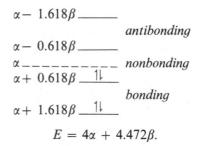

$$E = 4\alpha + 4.472\beta.$$

Note that conjugation stabilizes butadiene over two separate ethylene molecules by an additional 0.472β.

Once we have solved for the values of E, we can evaluate the coefficients in the four wave functions corresponding to the four energy levels, as we did for the H_2^+ case. The solutions are:

	c_1	c_2	c_3	c_4
ψ_{IV}	0.37	−0.60	0.60	−0.37
ψ_{III}	0.60	−0.37	−0.37	0.60
ψ_{II}	0.60	0.37	−0.37	−0.60
ψ_{I}	0.37	0.60	0.60	0.37

These represent the amplitudes of the individual atomic wave functions in the four MOs. These are illustrated below.

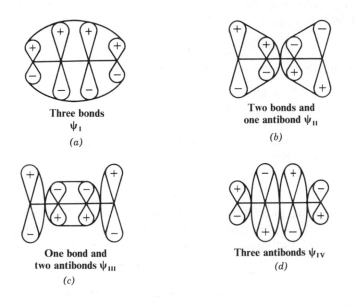

Three bonds
ψ_{I}
(a)

Two bonds and
one antibond ψ_{II}
(b)

One bond and
two antibonds ψ_{III}
(c)

Three antibonds ψ_{IV}
(d)

Where the signs of the overlapping lobes (as given by the signs of the co-efficients) are the same, a bond is formed, and where the signs change an antibond is formed. Thus the most stable wave function ψ_{I} consists of three bonds and no antibonds, and the second wave function ψ_{II} consists of two bonds and one antibond.

When the number of bonds is greater than the number of antibonds, we have a bonding energy level more stable than the unassociated atoms; when there are more antibonds than bonds, such as in ψ_{III}, we have an *antibonding* energy level. If the numbers are equal, we refer to this as a *nonbonding* level.

BOND ORDERS

Once we have calculated the coefficients of the wave functions, we can calculate the strength of the bonds by calculating the amount of π-bonding or bond order in each bond. The bond order p_{ij} is given by

$$p_{ij} = \sum_{1}^{\psi_{occupied}} Nc_i c_j, \tag{7.52}$$

It is the sum of the products of the coefficients between the two atoms making up the bond, summed over the occupied levels, where N is the number of electrons occupying that level. For butadiene, we can calculate p_{12} by recognizing that only ψ_I and ψ_{II} are occupied, each with two electrons. The coefficients between carbon 1 and carbon 2 in the lowest level are $+0.37$ and $+0.60$. They are $+0.60$ and $+0.37$ in the second energy level. Thus

$$p_{12} = 2(0.37)(0.60) + 2(0.60)(0.37) = 0.89$$

to two significant figures. In a similar fashion,

$$p_{23} = 2(0.60)(0.60) + 2(0.37)(-0.37) = 0.45$$

$$p_{34} = p_{12} = 0.89.$$

Since we have, in addition, a σ-bond between each carbon, we add 1.0 to each bond order, giving

$$\underset{1.89}{C} \text{—} \underset{1.45}{C} \text{—} \underset{1.89}{C} \text{—} C$$

We find a good linear relationship between these calculated bond orders and bond lengths. Using the bond lengths of 1.34 and 1.53 for ethylene and ethane, and the bond orders of 2.0 and 1.0, we can develop the equation

$$d_{ij} = (-0.19)p_{ij} + 1.72 \tag{7.53}$$

where d_{ij} is the bond length and p_{ij} is the calculated bond order. This equation predicts bond lengths of 1.44 and 1.36 Å for butadiene, which compare well with the measured butadiene bond lengths at 1.47 and 1.37 Å.

THE GENERAL PROCEDURE FOR WRITING DOWN THE MATRIX

We can make the following three generalizations for writing x's, 1's, and 0's in the determinant matrix for a given conjugated hydrocarbon:

1. All diagonal elements are x's.
2. All off-diagonal elements referring to connected atoms are 1's.
3. All other off-diagonal elements are 0.

Let us now calculate trimethylene methane as an MO system:

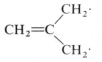

The carbons are numbered,

the matrix takes the form

$$\begin{vmatrix} x & 1 & 0 & 0 \\ 1 & x & 1 & 1 \\ 0 & 1 & x & 0 \\ 0 & 1 & 0 & x \end{vmatrix} = 0.$$

and the solution gives four energy levels. These are:

$$\alpha - 1.732\beta \quad \underline{\qquad}$$

$$\alpha \qquad \underline{\uparrow} \qquad \underline{\uparrow}$$

$$\alpha + 1.732\beta \quad \underline{\uparrow\downarrow}$$

which we fill with the four π-electrons, keeping those having the same energy unpaired, in accordance with Hund's rule. The α-level electrons are not stabilized by bonding and are termed *nonbonding* electrons. The bond orders are given below:

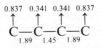

FREE VALENCE

The total bond order of the central carbon in trimethylene methane is $3(1.57735) = 4.732$. This is the maximum possible bond order for any carbon in a conjugated system, and any unused bonding capacity on carbons less bonded than this is referred to as the *free valence* F_i:

$$F_i = 4.732 - \sum p_{ij} \qquad \text{for all connected bonds.} \qquad (7.54)$$

For butadiene, the free valence is shown at the end of arrows along with the bond orders:

0.837 0.341 0.341 0.837
↑   ↑   ↑   ↑
C—C—C—C
  1.89 1.45 1.89

Free valences have been correlated with aromatic reactivities, and relative rates of reactivity toward free radical attack.

CHARGE DISTRIBUTION

The charge distribution is principally of value when dealing with charged species. It is given by:

$$q_i = 1.000 - \sum_{i=1}^{\psi_{occ}} N c_i^2 \qquad (7.55)$$

where N is the number of electrons in a given energy level and c_i is the coefficient for that atom in that wave function. For butadiene, the charge distribution is, of course, zero:

$$q_1 = q_4 = 1.000 - 2(0.37)^2 - 2(0.60)^2 = 0.000.$$

COMPUTER CALCULATION OF MO PARAMETERS

Obviously the factoring and solution of large determinants is a tedious job, and it is not generally done by hand. Instead, the simple Hückel MO calculations we have described here can easily be done using a computer program. There are several programs available from various sources that will perform these calculations. The program we will use here is called SHMO and has a long history documented in the comments of the program, shown in Appendix II. While originally a card-oriented program, this version has been modified for timesharing on the DEC system-10 so that only a few parameters are actually printed out at the terminal.

To use the program, you simply write down the matrix of x's, 1's, and 0's, which we have already discussed and enter its right-upper diagonal into the program. The SHMO program also allows for charged atoms and for heteroatoms to be included in the calculations. In the usual hydrocarbon case, the "core charges" are the net charge of each atom as seen by its π-electron and all have the value of 1.0. If charged species are used, a $+1$ carbon would have a core charge of 2.0 and a -1 charged atom a charge of 0.0.

To use the program, log into your system and start the program by typing RUN SHMO. The program will start by typing

SHMO
NAME OF MOLECULE:

Enter any name for your molecule, terminated by a Return. The next questions are:

NUMBER OF ATOMS =
NUMBER OF PI-ELECTRONS =

Answer these appropriately. Note that the number of π-electrons may or may not be equal to the number of atoms, depending on the system. For example, the cyclopentadienyl anion contains five atoms but six π-electrons. The program then asks for the core charges. For uncharged hydrocarbons these will all be 1's and they can all be entered on a single line separated by commas. If they are not all 1's, the values should be entered separated by commas in the same order as the atoms will be numbered in the matrix.

CORE CHARGES: 1,1,1,1,1,1   for example.

The program then asks the question OMEGA ITERATION? For SHMO calculations answer with an N followed by a Return. For charged species, answer with a Y and the program will iterate as described on page 222.

The last data to be entered is the matrix itself. Here the matrix differs from that used in the solution of the determinant by hand in that the diagonal elements, which are normally all x's, are entered as *zeroes* for hydrocarbons. Only the right-upper diagonal of the matrix is to be entered. For example, for benzene, the matrix is:

	1	2	3	4	5	6
1	x	1	0	0	0	1
2	1	x	1	0	0	0
3	0	1	x	1	0	0
4	0	0	1	x	1	0
5	0	0	0	1	x	1
6	1	0	0	0	1	x

for the molecule numbered

The upper-right diagonal is:

x	1	0	0	0	1	and the entries to the SHMO	1	0, 1, 0, 0, 0, 1
	x	1	0	0	0	program are	2	0, 1, 0, 0, 0
		x	1	0	0		3	0, 1, 0, 0
			x	1	0		4	0, 1, 0
				x	1		5	0, 1
					x		6	0

The program then calculates the energy levels, total energy (filling the levels from lowest to highest), the charge density, bond orders, and free valences. This is illustrated in Figure 7.1. When you have finished a calculation, the program will restart and ask for more cases. It can be stopped by entering a zero to the number-of-atoms question. The program will then ask if complete data are to be printed on the line printer. If the answer is N, no listing will be produced, if Y the coefficients and the complete matrix will be printed out.

```
SHMO
NAME OF MOLECULE: BENZENE
NO. OF ATOMS= 6
NO. OF PI-ELECTRONS= 6
CORE CHARGES: 1,1,1,1,1,1
OMEGA ITERATION?(Y OR N): N
ENTER UPPER DIAGONAL OF MATRIX
    1   0,1,0,0,0,1
    2   0,1,0,0,0
    3   0,1,0,0
    4   0,1,0
    5   0,1
    6   0
```

```
ENERGY LEVELS FOR BENZENE
ALPHA -  2.000 BETA   -----
ALPHA -  1.000 BETA   -----
ALPHA -  1.000 BETA   -----
ALPHA +  1.000 BETA   -----
ALPHA +  1.000 BETA   -----
ALPHA +  2.000 BETA   -----
```

```
            TOTAL ENERGY =   6 ALPHA +  8.00 BETA
```

```
ELECTRON DENSITY
    1   0.00
    2  -0.00
    3  -0.00
    4   0.00
    5   0.00
    6   0.00
```

```
BOND ORDERS
        1      2      3      4      5
    2   1.67
    3   0.00   1.67
    4  -0.33   0.00   1.67
    5  -0.00  -0.33  -0.00   1.67
    6   1.67  -0.00  -0.33   0.00   1.67
```

```
FREE VALENCES
    1   0.40
    2   0.40
    3   0.40
    4   0.40
    5   0.40
    6   0.40
```

```
SHMO
NAME OF MOLECULE:
NO. OF ATOMS= 0
PRINT OUTPUT ON LINE PRINTER?(Y OR N): N
STOP
```

Figure 7.1 Entries and responses to the computer program SHMO for calculation of the energy levels for benzene.

AROMATIC SPECIES

The calculation of benzene shows that the following energy levels are produced:

$$\alpha - 2\beta \quad \underline{\quad}$$
$$\alpha - \beta \quad \underline{\quad} \quad \underline{\quad}$$
$$\alpha + \beta \quad \underline{1\!\downarrow} \quad \underline{1\!\downarrow}$$
$$\alpha + 2\beta \quad \underline{1\!\downarrow} \quad .$$

Note that all middle levels are doubled or *degenerate*, so that four electrons will fill these levels before filling higher ones. This is the origin of Hückel's $4n + 2$ rule:

A planar monocyclic compound will exhibit aromatic properties if it has $4n + 2$ π-electrons, where n is any positive integer.

Thus Hückel systems include the cyclopropenium ion, the cyclobutadiene dianion, the cyclopentadienyl anion, benzene, the tropyllium ion, and the cyclooctatetraenyl dianion:

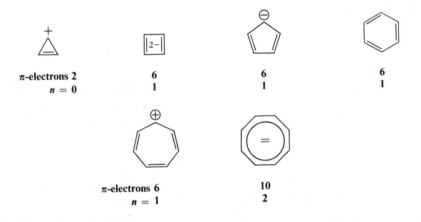

The energies of these systems can be calculated using the SHMO program, and resonance stability will be found for all of them.

NONAROMATICITY: CYCLOBUTADIENE

By contrast, the cyclobutadiene species is not a stable compound and has only been isolated at very low temperatures in an argon matrix.[1] The reason this compound has been so difficult to prepare can readily be determined from its energies. The energy levels are:

$$
\begin{array}{ll}
\alpha - 2\beta & \underline{\quad} \\
\alpha & \underline{1} \; \underline{1} \\
\alpha + 2\beta & \underline{1\!\!\downarrow} \quad .
\end{array}
$$

Adding up the energy of the system, we find that it is $4\alpha + 4\beta$, exactly the same energy as we find for two ethylene molecules. Thus we see that cyclobutadiene has no resonance stabilization energy over the two isolated ethylene molecules and we would not expect it to exist under normal conditions as a stable compound. In fact, the structure of the compound at these low temperatures has been reported not to be square at all, but rectangular, further justifying our belief that there is no significant π-overlap across the ring.

VARIATION OF α IN CHARGED COMPOUNDS

If the charge density of a particular C-atom is other than zero, then it is clear that α for that carbon is not the same as α_c. Coulombic attraction is increased if $q_i < 1$ since electron screening is reduced. This will reduce α. Wheland and Mann[2] proposed that when these charges are encountered, the α for that carbon should be varied to account for the charge according to

$$
\alpha_i = \alpha_c + (1 - q_i)\omega\beta_0. \tag{7.56}
$$

The value of α_i is varied through several iterations until self consistent q_i values are obtained. The constant ω has had a number of values, but is now usually taken as 1.40. The SHMO program has the capability of varying α's by this ω-technique by answering Y to the question OMEGA ITERATION?

CRITICISMS OF SHMO THEORY

We have seen that SHMO theory is a very powerful simple technique for predicting relative resonance energies and stabilities of molecules and for determining charge densities. In developing this treatment we have made a number of broad assumptions, which can be criticized. Unfortunately, the solution to many of the difficulties is a more complex theory.

1. *The wave functions may not be orthogonal.* We have assumed that there is no interaction between the atomic wave functions for each carbon π-orbital that we have mixed together to make the MO. In fact, there could be some overlap between them. This might occur particularly in strained or bent molecules.

2. *If the S_{ij}'s are different and nonzero, how can they be determined?* We next face the problem that if these overlap integrals are not zero, it is very difficult to evaluate what they might be. Furthermore, there are many conjugated systems containing heteroatoms that we would like to evaluate. We can be sure that the values of S_{ij} will vary when only one of the atoms is carbon.

3. *The β's may not all be the same.* For example, if the bond distance varies, as it does in various bonds in butadiene, the resonance integral may not be the same for all interactions.

4. *The α's will probably also vary in compounds in which heteroatoms are included.* Again, in heteroatom cases, we can be sure that the Coulomb integral will be different than that for carbon. This problem is dealt with in Table 7.1 below.

VARIATION OF α AND β

The value of the Coulomb integral α_x for a heteroatom X can be expressed by

$$\alpha_x = \alpha_c + h_x \beta$$

and the variation in β_{xy}, where X and Y may be heteroatoms as

$$\beta_{xy} = k_{xy} \beta_c.$$

The values for h_x and k_{xy} are given in Table 7.1 below.

Table 7.1   Parameters to Use in Varying α and β for Heteroatoms[3]

Atom or Group	h_x			k_{xy}		
	SHMO	OMEGA	Bond	SHMO	OMEGA	
C—N	0.5	0.8	C—C	0.9	0.7	
C=N	1.5	1.7	C≡C	1		(aromatic C—C)
R_4N^+	2	—	C—N	0.8	1.0	
C—O	1	1.3	C=N	1	0.7	
C=O	2	2.7	N—O	0.7	0.5	
R_3O^+	2.5	—	C—O	0.8	0.8	
F	3	—	C=O	1	0.6	
Cl	2	2.8	C—F	0.7	—	
Br	1.5	—	C—Cl	0.4	0.7	
			C—Br	0.3	—	

The parameters in Table 7.1 are used to modify the numbers entered in the matrix for the calculation of the energy levels by the SHMO method. In the case of hand calculation, the revised values of α and β are entered in the determinant. However, in the case of the computer calculation, the values of h_x and k_{xy} are entered in the matrix. For example, for SHMO calculation of furan, the matrix to be entered where O has position 1 is

1	1, 0.8, 0, 0, 0.8
2	0, 1, 0, 0
3	0, 1, 0
4	0, 1
5	0

SUMMARY

The SHMO method contains a fearsome number of approximations, but in spite of these it is remarkable that it correctly predicts resonance stabilization and aromaticity as well as correlating bond lengths with bond orders. In addition, charge densities measured by nmr chemical shifts have been correlated with the predicted charge densities from SHMO theory,[4] and as we will see in the problems that follow and in the next chapter, we can correlate IR bond stretching frequencies and UV transition frequencies with SHMO calculated parameters.

PROBLEMS

Problem 7.1

Calculate the resonance energies, bond energies, and free valances for the following molecules, using the SHMO program. Suggest reasons for any unusual bond orders or free valences that you find. Can they be correlated with chemical properties?

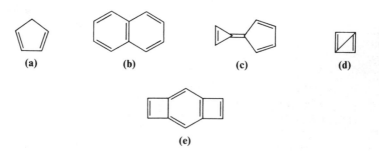

(a)   (b)   (c)   (d)

(e)

Problem 7.2

Explain why the cyclobutadienyl dianion is a Hückel system while cyclo-butadiene is so unstable that it cannot be prepared in ordinary ways.

Problem 7.3

Calculate the resonance energy for the allyl cation and anion.

Problem 7.4

The bond stretching wavelengths for some carbonyl compounds are given below. What SHMO parameter might they be correlated with?

CH_2O	5.73 μm
ϕCHO	5.85 μm
$\phi CO\phi$	6.00 μm
Napth—CO—ϕ	6.08 μm

RECOMMENDED READING

Coulson, C. A., O'Leary, B., and Mallion, R. B., *Hückel Theory for Organic Chemists*, Academic Press, London, 1978.

Dewar, M. J. S., *The Molecular Orbital Theory of Organic Chemistry*, McGraw-Hill, New York, 1969.

Dewar, M. J. S., and R. C. Dougherty, *The PMO Theory of Organic Chemistry*, Plenum Press, New York, 1975.

Kier, L. B., *Molecular Orbital Theory in Drug Research*, Academic Press, New York, 1971.

Orchin, M., and H. Jaffee, *Symmetry, Orbitals and Spectra*, Wiley-Interscience, New York, 1971.

Pople, J. A., *Approximate Molecular Orbital Theory*, McGraw-Hill, New York, 1971.

Roberts, J. D., *Notes on Molecular Orbital Calculations*, W. A. Benjamin, New York, 1962.

Salem, L., *The Molecular Orbital Theory of Conjugated Systems*, W. A. Benjamin, New York, 1966.

Streitwieser, A. Jr., *Molecular Orbital Theory for Organic Chemists*, Wiley, New York, 1961.

Turner, A. G., *Methods in Molecular Orbital Theory*, Prentice-Hall, Englewood Cliffs, N.J., 1975.

Woodward, R. B., and R. Hoffman, *The Conservation of Orbital Symmetry*, Academic Press, New York, 1970.

Yates, Keith, *Hückel Molecular Orbital Theory*, Academic Press, New York, 1978.

REFERENCES

1. S. Masamuna et al., *J. Chem. Soc. Chem. Commun.*, **1972**, 1268.
2. G. W. Wheland and D. E. Mann, *J. Chem. Phys.*, **17**, 264 (1949).
3. L. B. Kier, *Tetrahedron Lett.*, **1965**, 3273.
4. G. Fraenkel and J. W. Cooper, *Tetrahedron Lett.*, **1968**, 1825.

CHAPTER EIGHT

ULTRAVIOLET SPECTROSCOPY

In previous chapters, we discussed spectroscopy involving the stretching of bonds (IR) and nuclear interactions through electron-pair bonds (nmr). In this chapter we look at UV spectroscopy, which measures the energy absorbed when electrons are promoted into higher levels. Since this is an electron excitation phenomenon, UV is sometimes called electronic spectroscopy.

The measurement of UV absorption spectra is primarily used to detect the presence of conjugated hydrocarbons. The region we refer to as UV consists of the region from 1000–4000 Å or 100–400 nanometers (nm). While UV measurements are usually reported directly in nm, the corresponding range in cm^{-1} is 155,000–25,000 cm^{-1}.

We further subdivide the UV into the near UV and the far (or vacuum) UV as shown below:

800–400 nm	400–190 nm	190–100 nm	soft X-rays

Energy $\longrightarrow$

(red) visible (blue)	near UV	vacuum UV

The vacuum UV is so named because it is usually obscured by atmospheric absorptions and requires special vacuum equipment to observe it. Consequently, in this chapter we will be concerned almost exclusively with the near UV region.

Reviewing Beer's law (Chapter 1) we recall that

$$A = \varepsilon cb \tag{1.6}$$

where A = absorbance = $\log(I_0/I)$ = optical density

ε = molar absorptivity or extinction coefficient

c = concentration in moles/l

b = path length of sample cell in cm

While in IR spectra we were content with such labels as s, m, and w for peak intensity, we will report the value of ε in discussing UV absorptions. This molar absorptivity, ε, varies from 10^0–10^4; we can consider absorptions of

227

the order of 10^4 very strong and those less than 10^3 as weak. The peak intensity is a measure of transition probability, and peaks having low ε's often occur from transitions that are formally "forbidden." Since the ε-values vary over such a wide range, they are occasionally reported as logs.

An important corollary is that the observed absorbances are completely additive. If a sample contains two substances having ε_1 and ε_2 at a given λ then

$$A_T = b(\varepsilon_1 c_1 + \varepsilon_2 c_2).$$

We find that Beer's law holds fairly accurately for a wide range of compounds, except in the following cases:

1. When complexes form between the solvent and the compound, such as strong hydrogen bonding.
2. When the compounds show fluorescence or react photolytically.
3. When there are different forms of the absorbing system, such as an acid-base equilibrium.

In this last case, it is common to vary the pH of the solution and make several measurements. The absorption at some wavelengths will decrease in intensity and at other wavelengths will increase. A wavelength whose A does not change with pH is known as an *isosbestic point*. Such a point is usually the intersection between a peak due to the acid form and a peak due to the basic form.

SAMPLE HANDLING IN UV SPECTROSCOPY

The UV cell in various types of spectrophotometers is usually a quartz cell, but it might also be one made of fused silica, tridymite, or crystoballite. Quartz cells are useful only for wavelengths above 210 nm, but silica cells will allow observations of wavelengths as low as 165 nm when used in vacuum UV spectrometers. Solution cells can have path lengths from 1–10 cm, with 1 cm the most common. This 1-cm cell requires about 3 ml of solution, although filler plugs can be used to reduce the volume where necessary. Microcells with even smaller capacities are also available for extremely sample-limited measurements.

Typical solvents for use in UV spectroscopy are shown in Table 8.1, along with their cutoff frequency for 1-cm cells. Special, high purity spectral grade solvents from which UV-absorbing substances have been carefully removed must be used in nearly all cases. One major exception is that ordinary distilled water is usually sufficiently pure.

Table 8.1 Common Solvents Used in UV Spectroscopy

Solvent	Cutoff
Cyclohexane	190
Hexane	187
CCl_4	245
$CHCl_3$	223
CH_2Cl_2	215
Ethanol	198
Methanol	198
Water	197
Dioxane	215
Iso-octane	195

Unlike IR and nmr, the preparation of solutions for UV spectra requires great care in handling the sample and solvent, not only to prevent the introduction of unwanted substances, but so that the concentration will be known accurately. Obviously, considering Beer's law, we cannot know ε unless c has been carefully measured.

PRESENTATION OF SPECTRA

The plots produced by most spectrometers show absorbance A (or optical density, OD), versus wavelength. A more useful plot is made when this is converted to λ versus ε. Looking at the UV spectrum of acetone in Figure 8.1, we see that rather broad peaks are observed and that ε could be measured at any wavelength desired. By convention the molar absorptivity ε is reported at the maximum point on the curve. It should be noted that UV spectra are always plotted to show wavelength rather than frequency or wave number. This is perhaps unfortunate since either of the latter two cases would be linear in energy, but despite this objection all of the additivity rules that have been given for UV spectra are given in wavelengths. It is surprising that they work at all!

Let us consider the preparation of a sample for a UV spectrum in detail. If we are going to obtain a UV spectrum with a routine instrument, we need to calculate a concentration for the expected ε. We find that we can make the most accurate measurements if we can obtain an optical density or absorption

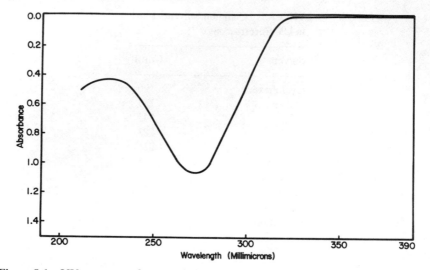

Figure 8.1 UV spectrum of acetone in iso-octane. In this presentation, the lowest point on the curve represents the maximum absorption. $\lambda_{max} = 274$, $\varepsilon = 17.6$.

of around 1.0, and we thus calculate a concentration based on the reported ε of alkyl-substituted unconjugated carbonyls of around 12, for our 1-cm cell.

$$c = \frac{A}{\varepsilon b}$$

$$c = \frac{1.0}{(12.1/\text{mole-cm})(1.0 \text{ cm})} = 0.0833 \text{ moles}/1.$$

For acetone (MW = 58) this corresponds to 4.83 g/1 or 0.120 g/25 ml. We then weigh about 0.12 g of acetone in a 25-ml volumetric flask and dilute it to 25 ml with a suitable solvent, in this case iso-octane. The spectrum is obtained using a matching solvent cell to balance the minor solvent absorptions, and ε is calculated at the point of greatest absorption, λ_{max}. For this spectrum of acetone

$$\lambda_{max} = 274 \text{ nm}$$

and

$$\varepsilon = \frac{A}{cb} = 16.7 \text{ l/mole-cm}.$$

SPECTROMETER DESIGN

Ultraviolet spectrometers have much the same design as IR spectrometers, with the major exception that they are totally enclosed to prevent stray light from striking the sensitive detectors. The source of light is usually a hydrogen discharge tube, which provides nearly a continuous band of radiation from 160–360 nm. Above 360 nm, in the transition to visible wavelengths, incandescent sources are used.

The detector in UV spectrometers is a sensitive photomultiplier tube. Such a tube operates by allowing the detected light to fall on a photocathode, where it dislodges a few electrons. The cathode is so arranged that the electrons will hit one of a series of *dynodes*, where each impinging electron causes several more to be expelled. As this occurs at each stage, the number of electrons after, say, nine stages of amplification can be considerable, and these electrons then strike the anode where the generated current can be used to drive a recorder. The output of this detector is transmittance (I/I_0), and if absorbance is desired [$\log (I/I_0)$], a logarithmic resistance is interposed between the detector output and the recorder.

VOCABULARY OF UV SPECTROSCOPY

Because UV spectroscopy has been in use for a long time, a number of unique terms are used to describe the spectral phenomena:

Chromophore—A chromophore is any functional group that absorbs UV light.

Bathochromic shift—In comparing the absorbance of one compound with another, we speak of a shift to the red or to longer wavelength and lower energy as a bathochromic shift.

Auxochrome—An auxochrome is an auxiliary functional group that interacts with a chromophore, causing a bathochromic shift.

Hypsochromic shift—This is the opposite of a bathochromic shift. A shift to shorter wavelength or toward the blue and higher energy is a hypsochromic shift.

Less frequently used terms are *hypochromic effect* and *hyperchromic effect*. A hyperchromic effect results in a compound having an enhanced ε compared to some reference compound, and a hypochromic effect results in a compound having a decreased ε with respect to some reference compound.

TRANSITIONS IN UV SPECTROSCOPY

When we looked at simple Hückel MO theory in the previous chapter, we found that considering only the π-electron levels, there were a number of *bonding* orbitals that were more stable than the unconjugated systems and *antibonding* orbitals which had an energy higher than that of the unconjugated systems. In addition, in a few cases we found levels exactly equal to α, which we termed *nonbonding*.

In the terminology of UV spectroscopy the bonding π-orbitals are termed π-levels and the antibonding levels π^*. By analogy, we find that there are also transitions between σ-bonding orbitals and higher energy antibonding levels called σ^*. We write these transitions as $\pi-\pi^*$ and $\sigma-\sigma^*$, and refer to them as "pi to pi-star" and "sigma to sigma-star" transitions. The $\sigma-\sigma^*$ transitions are all below 150 nm and are thus never observed directly. The $\pi-\pi^*$-transitions, on the other hand, are intense and are frequently observed in the near UV.

In addition to these two types of transitions, we observe transitions between unpaired electrons on heteroatoms such as oxygen, nitrogen, sulfur, and the halogens and antibonding orbitals. These nonbonding electrons are said to be in n orbitals and the transitions termed $n-\pi^*$ and $n-\sigma^*$. All of these transitions are "forbidden" because of the symmetry of the nonbonding orbitals compared to that of the σ^*- and π^*-orbitals and thus have ε's in the range of 100–3000. Some typical $n-\sigma^*$-transitions are shown in Table 8.2. Note that only a few are observed in the atmospheric UV.

Table 8.2 $n-\sigma^*$ Transitions

	λ(nm)	ε
CH_3OH	177	200
Bu_2S	210, 229	1200
$C_6H_{11}SH$	224	126
Me_3N	199	3950
CH_3Cl	173	200
n-PrBr	208	300
CH_3I	259	400

THE ETHYLENE CHROMOPHORE

The $\pi-\pi^*$-transition of ethylene is only about 165 nm, $\varepsilon = 10,000$, but alkyl substitution and conjugation with unshared electron pairs raise this wavelength to one in the near UV. These substituents, then, cause a *bathochromic*

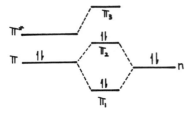

Figure 8.2 Combining of the n-, π-, and π^*-levels of ethylene and some chromophore with nonbonding electrons, producing a lower π_1 and a raised π_2. The distance between π_2 and π_3 is less than between the original π and π^*, thus causing a bathochromic (red) shift.

shift. The reason for this shift to lower energy can be explained by reference to Figure 8.2.

In this electron correlation diagram, we see on the outside the π- and π^*-orbitals of ethylene and an n-orbital from some atom providing an unshared pair of electrons. These energy levels interact to form higher and lower levels as shown, leading to two bonding orbitals called π_1 and π_2 and an antibonding orbital called π_3. The π_1-orbital is lower than the original π-orbital in energy, but the π_2-orbital is higher yet still a bonding orbital. Consequently, the distance from π_2 to π_3 is much less than that from π to π^*, and the energy required for the transition is less. This argument explains only the sharing of the electron pair as a resonance effect, but it can be expanded to include an inductive effect as well, with little change in the results.

Thus we see that substitution of this unshared pair causes a bathochromic shift, and it is thus an *auxochrome*. The argument is fairly plain for these heteroatoms, but such a bathochromic shift is also observed as alkyl substitution of the double bond occurs. We explain this similar shift by invoking hyperconjugation with the attached alkyl groups.

ABSORPTIONS IN DIENES

Once the ethylene chromophore becomes conjugated, the transition energies are much lower and result in absorptions in the atmospheric UV. This is more or less what we would have predicted from SHMO theory as we can see from the following. The energy levels of ethylene and butadiene are calculated to be:

$$\alpha - 1.618\beta \underline{\qquad}$$

$$\left.\begin{array}{l} \alpha - \beta \underline{\qquad} \\ \\ \alpha + \beta \underline{\qquad} \end{array}\right] \Delta E = 2\beta$$

$$\left.\begin{array}{l} \alpha - 0.618\beta \underline{\qquad} \\ \\ \alpha + 0.618\beta \; \underline{\text{⇅}} \end{array}\right] \Delta E = 1.236\beta.$$

ethylene

$$\alpha + 1.618\beta \; \underline{\text{⇅}}$$

butadiene

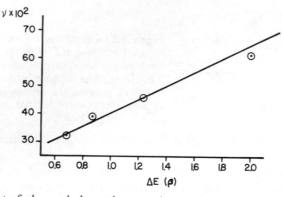

Figure 8.3 Plot of observed absorption energies (expressed as wave *numbers*) versus calculated SHMO ΔE's for some conjugated hydrocarbons H—(CH=CH)$_n$—H.

It is easy to see that the π–π^*-transition in ethylene requires an energy of 2β, but in butadiene only 1.236β, a lower energy and one more likely to be in the near UV.

Some actual measured wavelength maxima for conjugated systems are shown below:

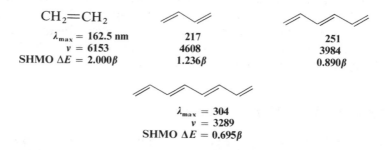

$$CH_2=CH_2$$

$\lambda_{max} = 162.5$ nm	217	251
$v = 6153$	4608	3984
SHMO $\Delta E = 2.000\beta$	1.236β	0.890β

$\lambda_{max} = 304$
$v = 3289$
SHMO $\Delta E = 0.695\beta$

Clearly, the higher the conjugation, the lower the energy of the transition. It is possible to correlate the predicted SHMO transition energy with that measured by UV absorption, as shown in Figure 8.3, where wave *numbers* (not wavelengths) are plotted against the calculated ΔE. The slope of this line allows us to calculate β as -60.5 Kcal.

WOODWARD'S RULES FOR DIENES

Woodward has reported a series of empirical rules for predicting the absorption maxima of dienes based on their substitution, conjugation, and double bond positions. These rules are:

Starting with 217 nm as the base of conjugated dienes,
1. Add 5 nm for each alkyl substituent.
2. Add 5 nm for each exocyclic double bond.

For example, we have the following cases of calculated and observed absorption maxima:

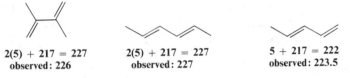

| 2(5) + 217 = 227 | 2(5) + 217 = 227 | 5 + 217 = 222 |
| observed: 226 | observed: 227 | observed: 223.5 |

In cyclic systems, we find that double bonds would "prefer" to be in the ring, and if exocyclic, their ground state is raised in energy, lowering their transition energy. This is the reason for Rule 2 and is illustrated below:

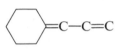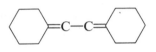

2 alkyl + 1 exo	3 alkyl	4 alkyl + 2 exo
3(5) + 217 = 232	3(5) + 217 = 232	6(5) + 217 = 247
observed: 236.5	observed: 235	observed: 248

WOODWARD AND FIESER RULES FOR CYCLIC CONJUGATED SYSTEMS

A more complete and useful set of rules has been developed by Woodward and Fieser[1] for conjugated cyclic dienes, notably those present in steroid systems. They start with either a homoannular or heteroannular diene base, depending on whether the double bonds are located in the same ring or in different rings:

Homoannular diene
base = 253 nm

Heteroannular diene
base = 214 nm

The set of additive rules are:

1. Base values
   (a) For homoannular diene start with a base of 253 nm.
   (b) For heteroannular diene, start with a base of 214 nm.
   (c) If both are present, use the homoannular base, 253 nm.
2. Additions for substituents
   (a) Add 5 for each alkyl group or ring residue.
   (b) If the alkyl group is attached to two double bonds, count it twice.
3. Add 30 for each additional double bond in conjugation.
4. Add 5 for each exocyclic double bond.

For example,

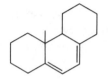 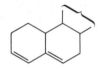

253 + 4(5) + 2(5) = 283   214 + 3(5) + 5 = 234
   alkyl   exo   alkyl exo

observed: 282, ε = 11,900   observed: 234, ε = 20,000

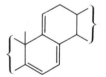

253 + 30 + 5(5) + 3(5) = 323
   db   alkyl   exo

observed: = 324, ε = 11,800

The following case shows an example where a single alkyl group must be counted twice, since the indicated carbon is a substituent on two different double bonds.

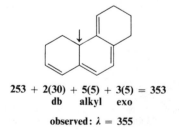

253 + 2(30) + 5(5) + 3(5) = 353
     db   alkyl   exo

observed: λ = 355

In cross-conjugated systems, we also use the longer (homoannular) base, and *ignore* the extra double bond in our calculation, since both cannot be in conjugation at once.

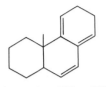

253 + 30 + 5(5) = 288
observed = 285, ε = 9100

Exceptions to these rules occur mainly when the double bonds are prevented from being co-planar because of some strain, notably in cases of adjacent exocyclic double bonds:

217 + 2(5) + 2(5) = 237
    exo   alkyl

but

217 + 2(5) + 2(5) = 237
    exo   alkyl

observed: 220, ε = 10,050

observed: λ = 243

The dimethylene cyclopentane presumably absorbs at the higher λ, because the five-membered ring allows greater co-planarity.

ABSORPTION OF CARBONYL GROUPS

Carbonyl groups have two types of readily available electrons that can undergo UV transitions: the π-electrons and the nonbonded unshared pair or n-electrons in an α-level on the oxygen. Since the nonbonded electrons are not stabilized directly by the formation of the π-bond, they are at a higher energy level than the π-electrons. Thus the distance to the π*-level is less for an n–π*-transition than for a π–π*-transition, and the n–π*-transitions are found at higher wavelengths (lower energy). This is shown in Figure 8.2. Since n–π*-transitions are forbidden by symmetry considerations, they have ε's on the order of 10–1000.

Conjugated systems such as α,β-unsaturated ketones show both an n–π*- and a π–π*-transition, both of lower energy (higher λ) than those of isolated systems. The nature of a transition in an unknown system may sometimes be evaluated by the effect of a change in solvent polarity on the wavelength of the transition. More polar solvents cause the $n–\pi^*$-transition to move toward a shorter λ (higher energy) and cause the $\pi–\pi^*$-transitions to move toward a longer λ (lower energy).

To explain these solvent effects, we consider the effect of hydrogen bonding on the ground state of the carbonyl chromophore:

$$\text{C=O}:\text{H—OR}$$

The nonbonded electrons on the oxygen will coordinate with hydroxylic solvents, lowering the net energy of the n-electrons. The distance to the π^*-level will then be higher, so that in a polar solvent the $n–\pi^*$-transition will be of higher energy or lower wavelength. The amount of shift to a lower wavelength is, in fact, a measure of the strength of the hydrogen bonding in that solvent.

The $\pi–\pi^*$-transition, on the other hand, shifts to a longer wavelength or lower energy in a more polar solvent. The $\pi–\pi^*$-transition has a polar excited state

$$\text{C=O} \xrightarrow{\ h\nu\ } \overset{\delta+\ \ \ \delta-}{\text{C---O}}$$

which would naturally be stabilized by hydrogen bonding in more polar solvents. This lowers the distance between π and π^*, thus lowering the energy or raising the wavelength of the transition. These effects are illustrated in Figure 8.4.

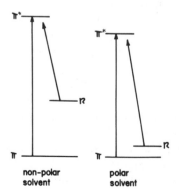

Figure 8.4 Solvent effects on $n–\pi^*$- and $\pi–\pi^*$-transitions. The $n–\pi^*$-transition is raised in energy in a more polar solvent, while the $\pi–\pi^*$-energy is lowered.

The solvent effect on the n–π^*- and π–π^*-transitions of 4-methyl-3-penten-2-one is shown in the following table:[2]

$$CH_3-\underset{\underset{CH_3}{|}}{C}=CH-\underset{\overset{O}{||}}{C}-CH_3$$

Solvent	$\pi - \pi^*$	ε	$n - \pi^*$	ε
Hexane	229.5	12,600	327	97.5
Ether	230	12,600	326	96
Ethanol	237	12,600	315	78
Methanol	238	10,700	312	74
H_2O	244.5	10,000	305	50

PREDICTION OF π-π^*-TRANSITIONS IN α, β-UNSATURATED KETONES

For cyclic unsaturated ketones, we can follow the three simple rules:

Parent α,β-unsaturated ketone	215
For each C-substituent α	10
β	12
γ and δ	18
Each exocyclic double bond	5
Each double bond extending conjugation	30

The total gives the predicted absorption in *alcohol*.

The following corrections are made to the π–π^*-transitions for solvent effects:

Alcohol	0
$CHCl_3$	1
Ether	7
Hexane	11
Water	-8
Dioxane	5

Some typical calculated compounds are shown below:

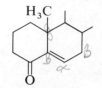
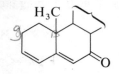

Δ^5-Androstene-17β-ol-4-one

215 + 10 + 12 + 5 = 242
base + α + β + exo
observed: 240

$\Delta^{3,5}$-Cholestadiene-7-one

215 + 12 + 18 + 30 + 5 = 280
base + β + δ + db + exo
observed: 280

AROMATIC COMPOUNDS

The UV spectra of benzenes are characterized by three major bands, which have been given the variety of names shown below:

187 nm	204 nm	256 nm
180 band	200 band	260 band
E_1	E_2	B
second primary	primary	secondary
β	p	α
^1B	^1L_a	^1L_b
$^1E_{2u}$	$^1B_{1\mu}$	$^1B_{2\mu}$
—	K	B

Only the 180 band is allowed, having $\varepsilon = 68{,}000$. The 204 band ($\varepsilon = 8800$) and the 256 band ($\varepsilon = 250$) have substantially lower intensities. The 204 band may appear more intense, however, since it may ride on the "tail" of the 184-nm band. The benzene UV spectrum has substantial fine struc-

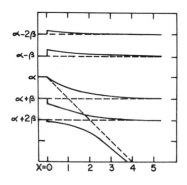

Figure 8.5 Change in energy levels for aromatic compounds as a function of the electronegativity of the substituent group. The n-level becomes the highest occupied MO and is then closer to a π^*-level, causing lower energy transitions (bathochromic shift).

ture because of interaction of the vibrational and electronic energy levels. This is illustrated in Figure 8.5.

Most substituents that can undergo conjugation with the ring, notably those with unshared pairs of electrons, cause a bathochromic shift of the benzene bands as shown below:[3]

Benzene	184 (68,000)	204 (8,800)	254 (250)
Toluene	189 (55,000)	208 (7,900)	262 (260)
Chlorobenzene		210 (7,500)	257 (170)
Bromobenzene		210 (7,500)	257 (170)
Aniline		230 (7,000)	280 (1,400)
			270
Phenol		211 (6,200)	258 (1,450)
Thiophenol		236 (10,000)	269 (700)
Anisole		217 (6,400)	269 (1,500)
N,N-Dimethylaniline		251 (14,000)	299 (2,100)
Nitrobenzene		252 (10,000)	280 (sh) (1,000)
			330 (sh) (140)

The bathochromic shift with increasing electronegativity is attributed to the fact that the Highest Occupied MO (HOMO) in the substituted benzenes is closer to the nonbonding α-energy and is thus closer to the π^*-level, decreasing the transition energy. This is illustrated in Figure 8.6 below. Note also that the primary and secondary bands shift to lower energy by a constant amount in most cases.

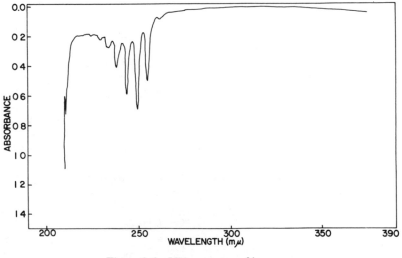

Figure 8.6   UV spectrum of benzene.

DISUBSTITUTED BENZENES

When there are no possible resonance interactions, disubstituted benzenes show a more or less additive effect of the two substituents on λ_{max}. The fine structure of the spectrum also disappears as more vibrational levels interact with the electronic ones. On the other hand, when resonance interactions can occur, the effects can be most striking. For example, the primary band of benzene is found at 204 nm, in chlorobenzene at 209.5 nm, and in benzoic acid at 230 nm. We would predict a λ_{max} for *p*-chlorobenzoic acid at 209.5–204 + 230 or 235.5 nm. We actually find it at 242 nm.

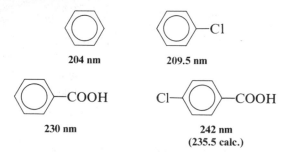

| 204 nm | 209.5 nm |

230 nm

242 nm
(235.5 calc.)

By contrast, where one of the groups is electron releasing and the other electron withdrawing by resonance mechanisms, we see a much greater bathochromic shift. We find the primary band in aniline at 230 nm, and in acetophenone at 245.5 nm. We would thus predict λ_{max} for *p*-acetylaniline at 230–204 nm + 245.5 or 271.5 nm. We actually find this absorption at 322 nm.

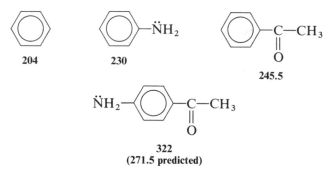

204   230

245.5

322
(271.5 predicted)

The great bathochromic shift is presumably due to resonance structures such as:

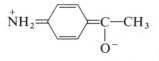

RESONANCE INTERACTIONS BETWEEN AROMATIC RINGS

The UV spectrum of biphenyl clearly indicates that some resonance inter-action occurs between rings: not only from the absorbance wavelength but also from the greatly enhanced molar absorptivity:

$\lambda_{max} = 204$, $\varepsilon = 7900$ $\lambda_{max} = 252$, $\varepsilon = 19,000$, primary bands

However, if the two aromatic rings are prevented from remaining co-planar so that maximum overlap cannot be achieved, the absorptivity is markedly diminished, as is illustrated for 2,2′-dimethylbiphenyl:

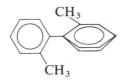

$\lambda_{max} = 270$, $\varepsilon = 800$, secondary band

Interposing a methylene group between the two rings as in diphenyl-methane produces a typical benzene-like spectrum:

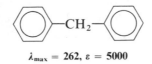

$\lambda_{max} = 262$, $\varepsilon = 5000$

However, if electron withdrawing and releasing groups are added to the two rings, some resonance interaction is still observed:

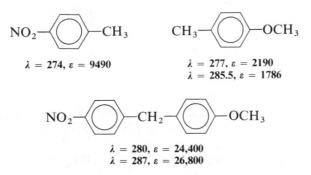

$\lambda = 274$, $\varepsilon = 9490$ $\lambda = 277$, $\varepsilon = 2190$
$\lambda = 285.5$, $\varepsilon = 1786$

$\lambda = 280$, $\varepsilon = 24,400$
$\lambda = 287$, $\varepsilon = 26,800$

CHARGE TRANSFER COMPLEXES

Charge transfer complexes occur between molecules that, when mixed, allow donation of electrons from one set of orbitals through space to another set. The classic example of this is the benzene-I_2 complex, where it is assumed that the benzene π-cloud interacts with the outer electrons of the iodine molecule. While both benzene and iodine have UV spectra of their own, a new band appears in the mixture at $\lambda = 290$ nm ($\varepsilon = 150{,}000$).

Although the mixture of compounds like 2,5-dimethoxy-*p*-xylene and

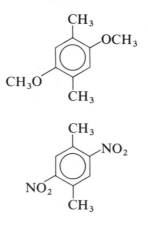

2,5-dinitro-*p*-xylene do not produce any new charge transfer bands, an extremely strong charge transfer complex has been observed between the two rings of the corresponding [2.2]-paracyclophane:[5]

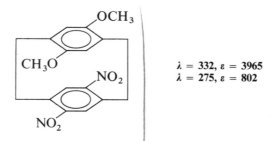

$\lambda = 332, \varepsilon = 3965$
$\lambda = 275, \varepsilon = 802$

In addition to having the indicated bands in the UV, this compound is also colored, indicating further transitions in the visible region:

$$\lambda = 468 \text{ nm } (4680 \text{ Å}), \varepsilon = 414.$$

PROBLEMS

Problem 8.1

Calculate the λ_{max} for the following compounds:

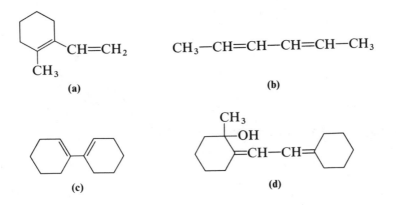

(a)

$$CH_3-CH=CH-CH=CH-CH_3$$

(b)

(c)

(d)

Problem 8.2

Calculate the λ_{max} for the following steroids:

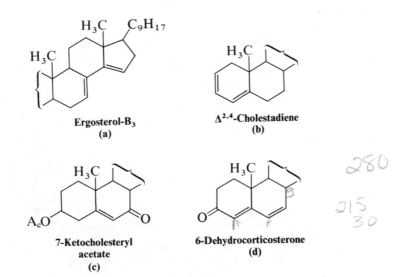

Ergosterol-B₃
(a)

$\Delta^{2,4}$-Cholestadiene
(b)

7-Ketocholesteryl
acetate
(c)

6-Dehydrocorticosterone
(d)

280

215
30

Problem 8.3

The enol acetate of cholestenone (**I**) could be either structure **II** or **III**. The observed λ_{max} is 238 (log ε = 4.2). Which structure is it? Assume that the acetate group has no substantial effect.

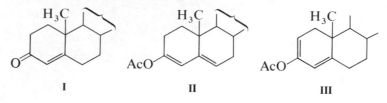

I   II   III

Problem 8.4

Ketones **IV** and **V** have been prepared and are observed to have λ_{max} = 241 and 247 nm. Which one has which λ_{max}?

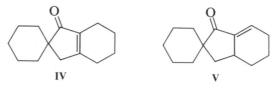

IV   V

Problem 8.5   UV λ_{max} = 226 nm; C_6H_{10}

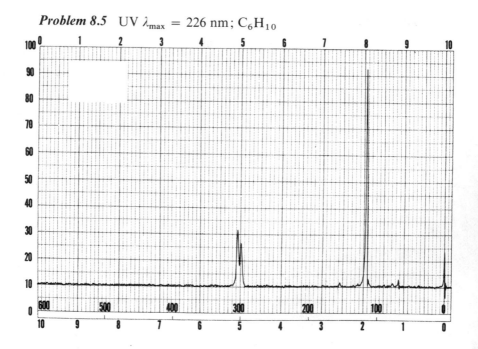

Problem 8.6   UV λ_{max} = 231 nm; C_7H_{12}

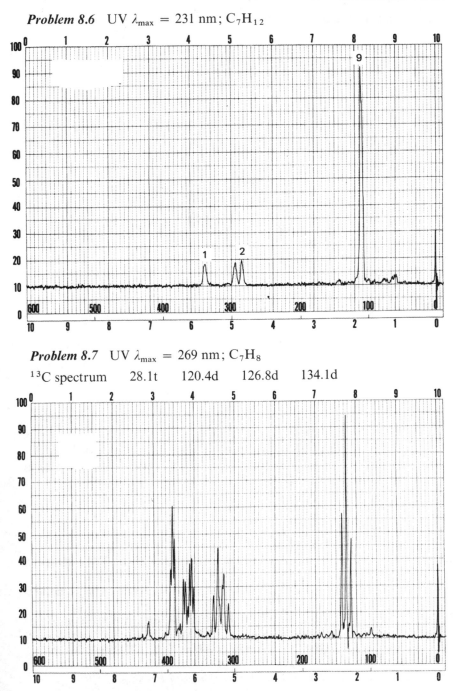

Problem 8.7   UV λ_{max} = 269 nm; C_7H_8

^{13}C spectrum   28.1t   120.4d   126.8d   134.1d

Problem 8.8 UV $\lambda_{max} = 235$, $\varepsilon = 14,000$; $C_6H_{10}O$

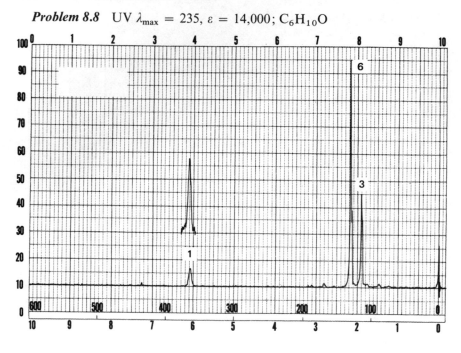

RECOMMENDED READING

Jaffe, H. H., and M. Orchin, *Theory and Applications of Ultraviolet Spectroscopy*, Wiley, New York, 1964.

Lambert, J. B., et al., *Organic Structural Analysis*, MacMillan, New York, 1976.

Pavia, D. L., G. M. Lampman, and G. S. Kriz, *Introduction to Spectroscopy*, W. B. Saunders, Philadelphia, 1979.

REFERENCES

1. L. F. Fieser and M. Fieser, *Natural Products Related to Phenanthrene*, Reinhold, New York, 1949.
2. G. J. Brealey and M. Kasha, *J. Am. Chem. Soc.*, **77**, 4462 (1955), as cited in Jaffee and Orchin, pp. 186–187.
3. J. R. Platt, *J. Chem. Phys.*, **17**, 484 (1949); E. P. Clar, *Aromatische Kohlenwasserstoffe*, Springer-Verlag, Berlin, 1952; L. Doub and J. M. Vanderbilt, *J. Am. Chem. Soc.*, **69**, 2714 (1947); **71**, 2424 (1949); A. Burawoy, *J. Chem. Soc.*, 1177 (1939); 20 (1941).
4. H. Suzuki, *Bull. Chem. Soc. Japan*, **32**, 1350 (1959).
5. H. A. Staab, C. P. Herz, and H.-E. Henke, *Tetrahedron Lett.*, 4393 (1974).

CHAPTER NINE

INTRODUCTION TO
MASS SPECTROSCOPY

The mass spectrometer is the first spectrometer we have discussed that actually destroys a small amount of the sample to obtain the spectrum. However, the amounts of material necessary to measure a mass spectrum are so small (as little as 1 μg) that we can, for all practical purposes, assume that no sample is destroyed. Since mass spectroscopy is so sensitive, we can often obtain information about extremely small amounts of materials where all other techniques would fail.

THE SPECTROMETER

The mass spectrometer is regarded by most chemists as being extremely complex, primarily because it usually requires a trained operator and operates under high vacuum. However, the principles involved are actually simpler than in, for example, nmr. A small amount of sample is introduced into an inlet port using a microsyringe and septum, a small bottle or break-off vial, or by touching a micropipette to a sintered glass opening. The sample chamber is maintained at fairly high vacuum (10^{-5} torr), and the sample is vaporized either at room temperature or with heating if the material is a solid. The vapor is allowed to slowly enter the ionization chamber through a small orifice called a molecular leak and is there ionized.

The ionization process most commonly consists of bombardment with high energy electrons:

$$M + e^- \text{ (fast)} \rightarrow M^{\pm} + 2e^- \text{ (slow)}$$

but can also be accomplished by UV light in some cases (photoionization), evaporation of material from a heated metal surface (surface emission), excitation by a spark in a vacuum chamber (vacuum spark source), release of ions near a surface by applying a strong electrical field (field ionization),

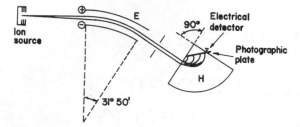

Figure 9.1 Simple diagram of a double-focusing mass analyzer (Mattauch-Herzog design). E is the electric field and H the magnetic field. Charged particles that are focused for a given m/e strike the detector or, alternatively, in older models, a photographic plate.

or charge exchange with previously ionized gas particles which then collide with the sample (chemical ionization).

The ions are then accelerated and focused using a combination of electrical and magnetic fields, similar to that as shown in Figure 9.1. In high resolution instruments there are two such fields, and the spectrometer is referred to as *double focusing*. The ions are accelerated through two slits to their final velocity and then scanned by varying either the magnetic or electrical field applied to them. The important point to recognize is that we are not observing molecular weights, but a ratio of mass to charge, or m/e. Since the charge is usually $+1$, this is most often proportional to mass, but if either the mass or the charge changes during acceleration and collection, the spectrum can be confusing, leading to broad "metastable" peaks and to doubly charged peaks at half the correct mass number.

For a given electrical and magnetic field, there will be only one m/e value that will strike the collector; the remaining masses will strike the sides of the tube instead. The collector amplifies the current produced by the striking ions and sends it to an oscilloscope, a plotter, or more commonly a minicomputer input.

The resolution of the mass spectrometer is given by the difference in mass that can be measured at a given mass size

$$R = \frac{M}{\Delta M}$$

where two peaks are said to be resolved if the valley between them is less than 10% of their height. For a medium resolution instrument, unit mass measurements are acceptable and the resolution need only be about 5000/1. High resolution instruments employing double focusing techniques have resolutions in the region of 10,000/1 to 100,000/1.

MEASUREMENT OF MASS

These very high resolution specifications imply that much more than unit mass measurements can be made meaningfully, and indeed the difference in mass between compounds containing different atoms but having the same nominal molecular weight can be detected. For example, the exact masses of several compounds having the nominal mass of 30 are shown in Table 9.1.

Table 9.1 Exact Masses of Some Fragments of Nominal Mass 30

NO	29.99800
CH_2O	30.01056
N_2H_2	30.02186
CH_2NH_2	30.03442
C_2H_6	30.04698

Table 9.2 Exact Masses of Common Isotopes

Element	Abundance	Mass
^1H	100	1.00783
^2H	0.016	2.01410
^{12}C	100	12.0000 (standard)
^{13}C	1.08	13.00336
^{14}N	100	14.0031
^{15}N	0.38	15.0001
^{16}O	100	15.9949
^{17}O	0.04	16.9991
^{18}O	0.20	17.9992
^{19}F	100	18.9984
^{28}Si	100	27.9769
^{29}Si	5.10	28.9765
^{30}Si	3.35	29.9738
^{31}P	100	30.9738
^{32}S	100	31.9721
^{33}S	0.78	32.9715
^{34}S	4.40	33.9679
^{35}Cl	100	34.9689
^{37}Cl	32.5	36.9659
^{79}Br	100	78.9183
^{81}Br	98	80.9163
^{127}I	100	126.9045

Since atomic masses are not, in fact, unit masses, but vary somewhat because of the binding energy of various atoms, we can identify the elemental composition of any peak from an accurate measurement of its mass. The masses of common isotopes are given in Table 9.2.

ELEMENTAL COMPOSITIONS IN UNIT MASS MEASUREMENTS

It is not always necessary, however, to obtain exact mass measurements in order to determine elemental composition. The abundance of isotopes other than the common ones in ordinary compounds allows us to estimate the elemental composition of a compound from the molecular ion M^+ and the $M + 1^+$ and $M + 2^+$ peaks which are due to these other isotopes. For example, in the spectrum of benzene shown in Figure 9.2, we see the peak of the *parent ion* or *molecular ion* M^+ at 78 and another peak 6.6% as high at 79 amu. Since there are six carbon atoms in benzene, each having a 1.09% probability of being ^{13}C, we would expect the M + 1 peak to be 6 × 1.09

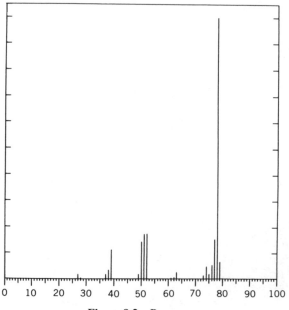

Figure 9.2 Benzene.

or about 6.5% as high as the M peak. This is indeed the case, and we find that this is often a fair predictor of the number of carbons in a molecule.

The intensity of the M + 1 peak, unfortunately, is not always so accurate, since carbon varies somewhat in isotopic composition. Furthermore, the M + 1 peak may be larger than expected because of ion–molecule reactions

$$M^{\ddagger} + MH \rightarrow MH^{+} + M\cdot$$

and because of contributions from the isotopes of other atoms.

This technique can also be used with caution to calculate the number of oxygens in a molecule. For example, the mass spectrum of acetic acid shows the parent ion at 60 and peaks at 61 and 62 having the ratios shown below:

60	100
61	2.52
62	0.47

Since we see around 2.2% in M + 1 we might reasonably predict that the molecule contains two carbons. The number is higher than that because of the contribution of 2(0.04) for ^{17}O. We suspect two oxygens because of the M + 2 peak of 0.47. The contribution of 0.20 from ^{18}O gives us an estimate of two oxygens immediately. The number is a little high (0.011 times 2.52) because of contributions from molecules containing two ^{13}C atoms. We predict then:

$$M + 1 = 2(1.1) + 2(0.04) = 2.28 \ (2.52 \ observed)$$

$$M + 2 = 2(0.20) + 2(0.011)(2.52) = 0.46 \ (0.47 \ observed).$$

These calculations can become even more complex when there are strong M − 1 peaks to include as well as N, O, and S. Since such calculations are seldom very accurate, no closed form of an equation is presented, particularly since it can have an extremely large number of terms if all of these peaks and isotopes are included.

Instead, the empirical scheme described by Biemann, in which the elements present are grouped into three classes, is often used:

1. Elements with low abundance of heavy isotopes, but which may be present in large quantities: C, H, N, and O.
2. Elements with heavy isotopes of high abundance but occurring in smaller numbers: S, Cl, Br, and Si.
3. Monoisotopic elements: F, P, and I.

These groups can be treated separately and the results added together. In addition, we can add the C and H atoms together and arrive at:

$$1.09 \times \text{ the number of C atoms}$$

or

$$1.10 \times \text{ the number of CH groups}$$

or

$$1.12 \times \text{ the number of CH}_2 \text{ groups.}$$

This method is useful for a rough estimation of the isotopic composition.

In the case of the halogens, particularly bromine, there are two isotopes having a substantial natural abundance that can lead to some rather distinctive patterns near the M^+ peak. This is illustrated in the spectra of methyl bromide and 1,1-dibromoethane shown in Figures 9.3 and 9.4. This doublet and triplet pattern is typical of brominated compounds and is readily recognized. Note that brominated compounds show peaks 2 amu apart, while chlorinated compounds show peaks 1 amu apart. We might note that talking about the molecular weight of a compound is somewhat ambiguous when there are substantial numbers of molecules having two

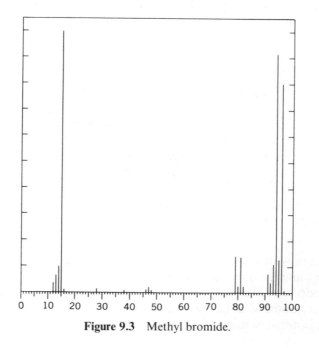

Figure 9.3 Methyl bromide.

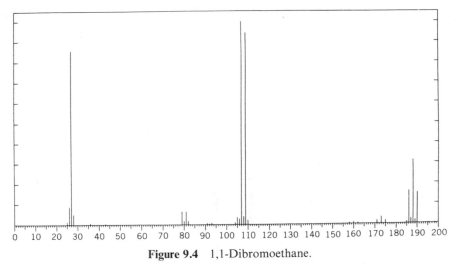

Figure 9.4   1,1-Dibromoethane.

molecular weights. By convention we always speak of the lowest molecular weight as being the true one and those of higher weight as being isotopes, even though in the case of atoms like bromine the abundances of the two isotopes are virtually identical.

RECOGNIZING THE PARENT ION M⁺

Thus far we have implied that the highest peak of substantial intensity will be the parent ion, but this is not always so. Even in the case of acetic acid, the largest peak is not the parent ion, but is $m/e = 43$, resulting from loss of OH from the molecule. (In some cases the parent ion is so small that it is virtually invisible, because of the tendency of the parent ion to fragment to smaller pieces.) While these small pieces contain a great deal of information about the structure of the compound, they are of little use if we cannot determine what its overall size is.

The probability of fragmentation occurring so efficiently that the parent ion will not have substantial intensity increases in the following series:

Aromatic compounds < conjugated alkenes < alicyclic compounds < sulfides < straight chain hydrocarbons < thiols < ketones < amines < esters < ethers < carboxylic acids < branched hydrocarbons < alcohols.

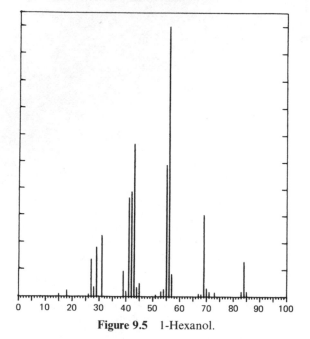

Figure 9.5 1-Hexanol.

By the time we get to highly branched hydrocarbons and alcohols we may not see the parent ion at all. In hydrocarbons, the fragments lost are multiples of $(CH_2)_n$, and these fragments may sometimes be put back together by examining the entire mass spectrum. Alcohols invariably lose H_2O, and if some other technique such as IR or nmr suggests an alcohol, the operator should be instructed to look carefully for a peak 18 amu above the last substantial peak, while varying the ionizing voltage.

The spectrum of 1-hexanol, for example, in Figure 9.5, shows a strong peak at 84 amu, but substantial peaks at both 83 and 85 amu, which is unlikely for a C, H, O compound. Careful examination of the region at $84 + 18 = 102$ amu with higher amplifier gain will show a very small peak, only 0.12% of the largest peak, at $m/e = 43$.

MULTIPLY CHARGED AND METASTABLE IONS

Since the mass spectrum is really a plot of m/e versus abundance rather than mass, there is the distinct possibility that doubly charged ions may occur in some species. For example, in the hexanol spectrum in Figure 9.5 we see that there is a small peak at m/e 51. There is no easy way to assemble C, H, and O to arrive at this mass, and the peak is probably a doubly charged hexanol ion:

$$CH_3(CH_2)_4CH_2OH^{++}$$

$m = 102, e = 2, m/e = 51$

These doubly charged ions are occasionally useful when the parent ion is extremely small, as it is in hexanol.

Such double charged ions may be recognized in two ways. First, if the compound itself has an odd mass, the doubly charged ion will appear at a half-integer mass. Second, such doubly charged ions require more energy to generate and usually do not appear in mass spectra where the bombarding electrons have less energy than about 30 eV. If the peak is of substantial intensity and at an integer mass, we would similarly expect a peak at the next half-integer mass to represent those molecules containing one ^{13}C atom.

There is also another class of peaks that appears at noninteger masses, classed as *metastable peaks*. These are usually quite broad and weak in intensity. They occur when a mass m_1 is accelerated and then begins to decompose:

$$m_1^+ \longrightarrow m_2^+ + R\cdot$$

The neutral fragment $R\cdot$ will continue with some of the kinetic energy, but the mass m_2 that remains charged will be accelerated and deflected. The result is a peak having properties of m_1 and m_2 but occurring at a new mass m^*, which is given by

$$m^* = \frac{(m_2)^2}{m_1}$$

where $m_1 > m_2$. These metastable peaks are not shown specifically in any spectra in this book because of their extremely low intensity. Recognition of these peaks can, however, help in determining both the original mass and the size of the ejected fragment. Such metastable peaks are observed primarily when the lifetime of the original species m_1 is in the range of 10^{-4}–10^{-6} sec: long enough to reach the accelerating region, but not long enough to become fully accelerated.

For example, a metastable peak at m/e 92.1 is observed in the spectrum of acetophenone (m/e 120) corresponding to the fragmentation

$$C_6H_5COCH_3^+ \longrightarrow C_6H_5CO^+ + CH_3\cdot$$

m/e 120 m/e 105

and the predicted mass m^* is

$$m^* = \frac{(105)^2}{120} = 91.88.$$

Unfortunately, many modern computer-controlled mass spectrometers do not plot these metastable peaks under routine conditions and the information inherent in them is thus ignored.

THE NITROGEN RULE

Finally, we can make the following generalization about the parent ion for compounds containing nitrogen:

> Only compounds having an odd number of nitrogens will have a parent ion at an odd mass number. Those with even numbers of nitrogens (including 0) will always have parent ions at even mass numbers.

This occurs, of course, because nitrogen has an odd number of substituents but an even mass number, while carbon, oxygen, and sulfur have even mass numbers and even valences, and the halogens and phosphorus have odd mass numbers and odd valences. This rule can also be extended to certain types of fragments.

FRAGMENTATION

Let us consider exactly what is happening in the ionization chamber of the mass spectrometer. A high-energy electron beam bombards the vaporized molecules and knocks out one electron:

$$M + e \text{ (fast)} \rightarrow M^{\ddagger} + 2e \text{ (slow)}.$$

The result is the parent ion $M^{\ddagger}$ we have been discussing. But this is not a carbonium M^+ containing a trivalent carbon less one-electron pair bond; instead it is the original molecule with one electron removed. The result is a *radical cation*, symbolized more accurately by $M^{\ddagger}$, which somewhere contains a *one-electron bond* or has a nonbonding electron removed from a heteroatom. Thus the molecule is no longer as stable and will tend to break apart. It is this breaking apart that we study in mass spectroscopy as a powerful structural tool.

There are two major types of fragmentation: elimination of a neutral molecule and elimination of a free radical. Neutral molecules such as ethylene, water, CO, HCl, HCN, and H_2S are preferentially eliminated in a number of fragmentations:

$$CH_3(CH_2)_3CH_2CH_2OH^{\ddagger} \longrightarrow CH_3(CH_2)_3CH{=}CH_2^{\ddagger} + H_2O$$

$$R_2CHCH_2CH_3^{\ddagger} \longrightarrow R_2CH^{\ddagger} + CH_2{=}CH_2$$

$$CH_3CH_2CN^{\ddagger} \longrightarrow CH_2{=}CH_2^{\ddagger} + HCN$$

Just as common is the separation of the ion part and the radical part into two fragments:

$$R_2CH\text{—}CH_2R^{\ddagger} \longrightarrow R_2CH^+ + RCH_2^{\cdot}$$

$$\underset{\substack{\| \\ O}}{R\text{—}C\text{—}R^{\ddagger}} \longrightarrow R\text{—}C{\equiv}O^+ + R\cdot$$

These two types of fragmentation differ in an important respect. Elimination of neutral molecules will produce a new fragment whose mass number has the same parity (odd or evenness) as the parent, while elimination of free radicals will change that parity. The new radical cations formed by elimination of neutral fragments are termed *odd electron* fragments, and those formed by elimination of a free radical are termed *even electron* fragments. This rule will allow us to decide on the type of fragmentation process that takes place in many cases.

Let us consider Example 1 in Figure 9.6. This compound has large peaks at 122 and 124, which suggest the presence of one bromine atom. Subtracting the atomic weight of bromine (122 − 79 = 43) we would predict a fragment having a mass of 43. There is indeed a large peak at this mass, which we might suggest is caused by the saturated alkyl group C_3H_7. This suggestion

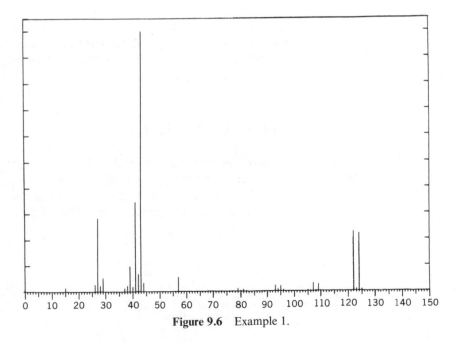

Figure 9.6 Example 1.

is confirmed by the fact that the M + 1 peak (123) is about 3.3% of the M^+ parent ion, thus indicating that three carbons are present in the molecule. We can then begin identifying major fragments. Clearly,

$$C_3H_7Br^{\ddagger} \longrightarrow C_3H_7^+ + Br\cdot$$

$$m/e\ 122 \qquad\qquad m/e\ 43$$

and indeed m/e 43 is the strongest peak in the spectrum. Note that the elimination of the Br· radical from the radical cation has generated an even electron fragment at m/e 43 having an *odd* mass, while the parent radical cation was an odd electron fragment having an even mass.

Another fragmentation reaction that generates an even electron product (having an odd mass) is:

$$C_3H_7Br^{\ddagger} \longrightarrow CH_2Br^+ + CH_3CH_2\cdot$$

$$m/e = 122 \qquad\qquad m/e = 93$$

In this case, the presence of this peak for the most part defines the parent compound as being a primary alkyl bromide, 1-bromopropane.

A fragmentation reaction leading to no change in parity of mass and which is also somewhat indicative of the structure is

$$CH_3{-}CH_2{-}CH_2Br^{\ddagger} \longrightarrow CH_3{-}CH{=}CH_2^{\ddagger} + HBr$$

$$m/e = 122 \qquad\qquad\qquad m/e = 42$$

This is a more prominent peak in the 1-bromopropane spectrum than in that for 2-bromopropane, since this particular fragmentation seems to occur more favorably when the hydrogens are available from carbons three or more bonds distant from the Br atom.

GENERAL FRAGMENTATION REACTIONS

All of the common cleavage, rearrangement, and elimination reactions that take place in the mass spectrometer have been grouped into 13 categories, termed A_1–A_5, B, C, D, E_1, E_2, F, G, and H. We will present them slightly out of alphabetical order to emphasize the fundamental similarities among a number of these types.

Type A_1

$$-\overset{|}{\underset{|}{C}}-\overset{|}{\underset{|}{C}}-^{\ddagger} \xrightarrow{\ A_1\ } -\overset{|}{\underset{|}{C}}^+ + \cdot\overset{|}{\underset{|}{C}}-$$

This is the standard hydrocarbon cleavage reaction, which decomposes the odd electron radical cation into an even electron cation and a neutral radical. Rearrangement of the product cation may, of course, occur and is not always detectable. The spectrum of Example 2 in Figure 9.7 has a parent ion peak at 86 amu and the first strong peak at 71 or M-15. Since methyl groups are seldom observed to leave straight chain compounds, this implies a branched methyl group. The peak at 72 has an intensity 0.054 times that at 71, implying a carbon–hydrogen chain of five carbons. Thereafter, the peaks at 71, 57, 43, 29, and 15 form a *homologous series* resulting from successive loss of CH_2 fragments. The compound is a methylpentane. The position of the alkyl branch can be deduced by considering the possible fragments that might occur by type A_1 cleavage:

$$CH_3CH_2CH_2CHCH_3 \xrightarrow{A_1} CH_3CH_2CH_2\overset{+}{C}HCH_3 + CH_3\cdot$$

with $\underset{CH_3}{|}$ below the fourth carbon, labeled $m/e\ 84$; product labeled $m/e\ 71$

$$\xrightarrow{A_1} CH_3-CH_2CH_2\cdot + CH_3\overset{+}{C}HCH_3$$

$m/e\ 43$

versus

$$CH_3CH_2CHCH_2CH_3 \xrightarrow{A_1} CH_3CH_2\overset{+}{C}HCH_3 + \dot{C}H_2CH_3$$

with $\underset{CH_3}{|}$ below the third carbon; product labeled $m/e\ 57$

$$\xrightarrow{A_1} CH_3CH_2\overset{+}{C}HCH_2CH_3 + CH_3\cdot$$

$m/e\ 71$

Since m/e 71 and 43 have substantially greater intensities than m/e 57, the compound is identified as 2-methylpentane.

Type A_5

$$RX^{\ddagger} \xrightarrow{A_5} R^+ + X\cdot$$

This type is characteristic of alkyl halides, but is also observed where $X = OR, SR,$ or NR_2. We have already seen this pattern in methyl bromide. Example 3 in Figure 9.8 shows a parent ion at 102 amu. Since it is even, it contains no nitrogens or an even number of nitrogens. The ratio between

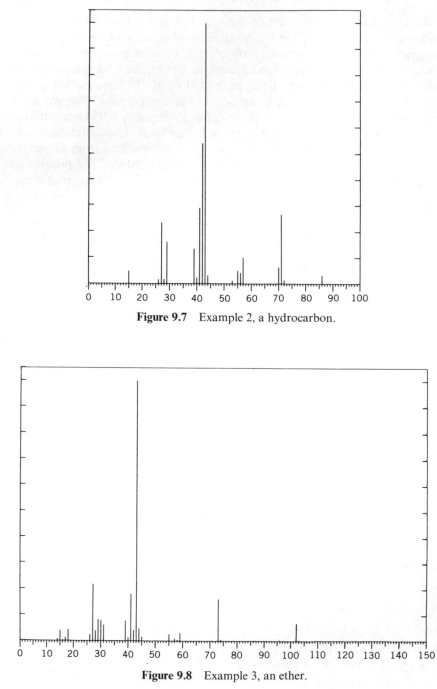

Figure 9.7 Example 2, a hydrocarbon.

Figure 9.8 Example 3, an ether.

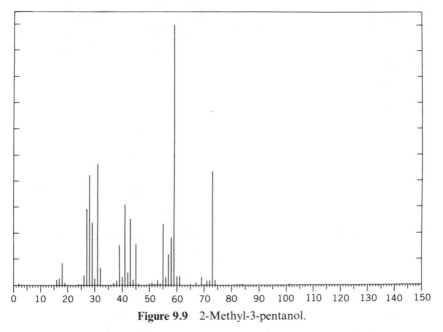

Figure 9.9   2-Methyl-3-pentanol.

102 and 103 (M + 1) corresponds to $C_6H_{14}O$, and since there is no M-18 peak, the compound is an ether. The major fragment at m/e 43 corresponds to

$$C_3H_7OC_3H_7 \cdot^{+} \xrightarrow{\;\;A_5\;\;} C_3H_7^{+} + \dot{O}C_3H_7$$
$$m/e\ 102 \qquad\qquad m/e\ 43$$

by the A_5 cleavage process. Since there is no major peak at $43 - 15 = 28$, the chain is not branched and the compound is *n*-propyl ether. In asymmetrical ethers, there would be some contribution by the cations from each side chain attached to the oxygen.

Type *B*

Cleavage types A_1 and A_5 are quite similar in that a carbon-X bond is ruptured, leaving a cation and a radical. In type B cleavage, the carbon–carbon bond adjacent to a *heteroatom* is ruptured, leaving the heteroatom to stabilize the positive charge:

$$\underset{\underset{\textstyle X}{|}}{R_2-C-R}\cdot^{+} \xrightarrow{\;\;B\;\;} \underset{\underset{\textstyle X}{|}}{R_2-\overset{+}{C}} + R\cdot \longrightarrow R_2-C{=}\overset{+}{X}$$

This fragmentation is characteristic of alcohols, esters, sulfides, thiols, and halogens. In Figure 9.9, the spectrum of 2-methyl-3-pentanol has a parent ion of 102, which is barely visible and major fragments at 73 and 59 corresponding to fragmentation patterns:

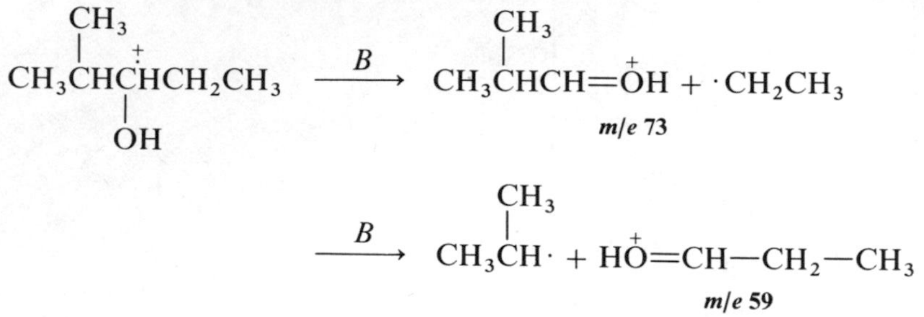

$$CH_3CHCHCH_2CH_3 \xrightarrow{B} CH_3CHCH=\overset{+}{O}H + \cdot CH_2CH_3$$

$m/e\ 73$

$$\xrightarrow{B} CH_3\dot{C}H\cdot + H\overset{+}{O}=CH-CH_2-CH_3$$

$m/e\ 59$

Note that for this branched secondary alcohol, these fragments are so stable that there is virtually no M-18 peak at all.

Type C

This fragmentation route is quite like type B except that it occurs at bonds adjacent to carbonyl groups in ketones, esters, and amides:

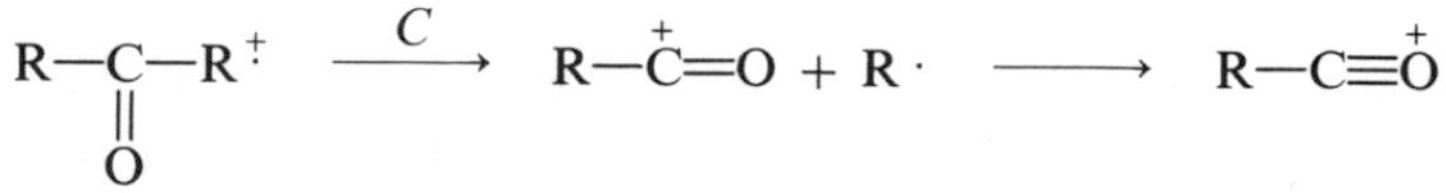

$$R-C-R\overset{+}{\cdot} \xrightarrow{C} R-\overset{+}{C}=O + R\cdot \longrightarrow R-C\equiv\overset{+}{O}$$

In the spectrum in Figure 9.10, the major fragments are 57, 43, and 29, with an important fragment at 71. The peak at m/e 43 is nearly always $C_3H_7^+$ and that at m/e 29 is $C_2H_5^+$. There is a small M-15 peak at 85, indicating a branched methyl group, and we can easily show by subtraction that this is most probably a ketone ($100 - 29 - 43 = 28$). The large peak at 57 amu is due to

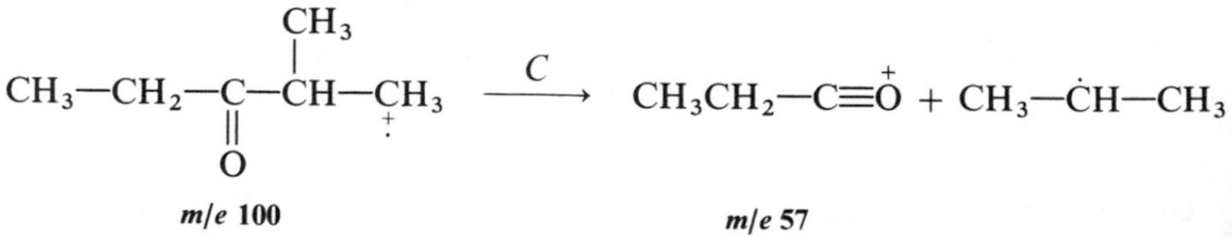

$$CH_3-CH_2-C-CH-CH_3 \xrightarrow{C} CH_3CH_2-C\equiv\overset{+}{O} + CH_3-\dot{C}H-CH_3$$

$m/e\ 100$ $m/e\ 57$

and that at 71 is due to

$$\xrightarrow{C} \overset{+}{O}\equiv C-CH-CH_3 + \dot{C}H_2-CH_3$$

$m/e\ 71$

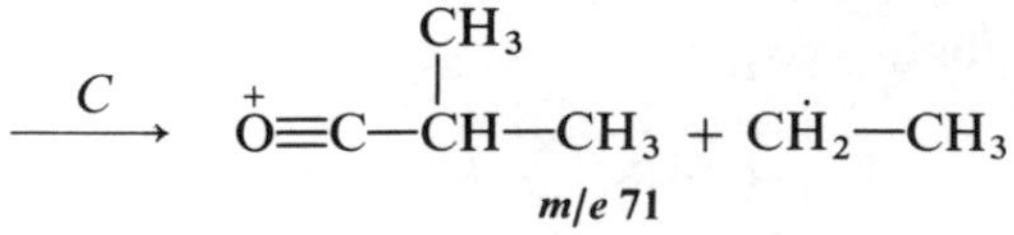

The compound is 2-methyl-3-pentanone.

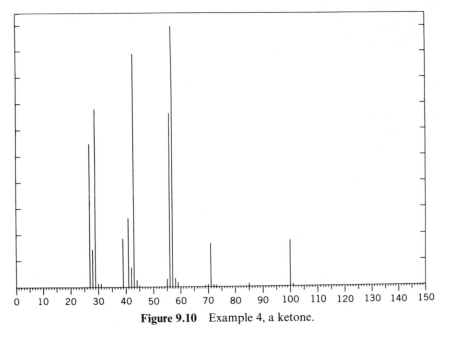

Figure 9.10 Example 4, a ketone.

Type A_2

The last of the simple single bond cleavages is type A_2. It involves the motion of an electron pair and the elimination of a neutral molecule, such as ethylene:

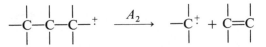

It has been shown by labeling experiments to be of little importance.

CLEAVAGE NEAR DOUBLE BONDS

Types A_3, A_4, and D are cleavages that occur near double bonds and other conjugated systems.

Type A_3

$$C{=}C{-}C{-}C{-}\overset{+}{\cdot} \xrightarrow{\quad A_3 \quad} C{=}C{-}C^+ + \cdot\, C{-}$$

Cleavage of a single bond β to a double bond is facilitated by the stabilization of the developing allyl cation:

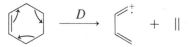

This does not make the identification of double bond placement any easier, however, because double bonds in M^+ migrate rather easily, so that the cleavage may occur in any number of places along the chain and still be an allylic cation. In cyclic systems, the double bonds are less mobile and these fragmentations can be more useful.

Type *D*

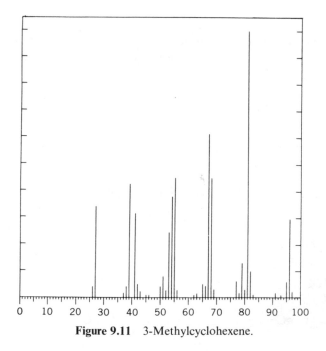

In cyclic unsaturated systems, a retro-Diels–Alder reaction can occur, leading to a stable charged diene fragment and an uncharged ethylene fragment. These fragments can often be used to deduce the position of the double bond, unlike the case of straight chain compounds and the A_3 cleavage above. For example, the 3-methylcyclohexene mass spectrum in Figure 9.11 shows the

Figure 9.11   3-Methylcyclohexene.

following pattern:

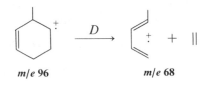

m/e 96 $\qquad\qquad$ m/e 68

Note that when the neutral molecule ethylene is eliminated, the mass fragment remaining (m/e 68) occurs at an *even* mass, having the same parity as M^+. All of the previous fragmentation types have resulted in odd mass fragments.

Type A_4

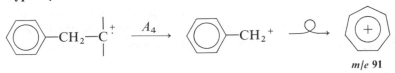

m/e 91

While it would be expected that the benzylic cation would be relatively stable, it has been shown that the fragment at m/e 91 for ethylbenzene is actually in equilibrium with the tropyllium ion, which because of its aromaticity is formed by rearrangement. Furthermore, while we would not expect a methyl group to be lost easily from an aromatic ring such as toluene, the xylenes (dimethylbenzenes) easily lose a methyl group and exhibit the characteristic tropyllium peak at m/e 91.

REARRANGEMENTS

The rearrangement fragmentation pathways all generate a neutral fragment and a new odd electron radical cation M^+. In these cases a bond is broken that might not in itself be energetically favorable, but which coupled with the transfer of a nearby hydrogen leads to a favorable fragmentation. Molecules lacking bonds that are easily cleaved are therefore likely to undergo these rearrangement fragmentations. Neutral molecules are eliminated in all cases, and the major fragments will have the same parity as M^+.

Type E_1

The general form of this elimination

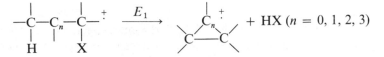

obscures the fact that the charged product may be an olefin. This is the general mechanism for the elimination of water from alcohols. In simple alcohols, the product is an *alkene*, the "two-membered ring" where $n = 0$.

$$RCH_2CH_2OH \overset{+}{\cdot} \xrightarrow{E_1} R-CH=CH_2 \overset{+}{\cdot} + H_2O$$

However, even in small alcohols, the product may not always be an alkene, but rather it may be a small ring. For example, deuterium labeling of 1-butanol has shown that the dominant elimination mechanism is 1,3 and 1,4, suggesting that the product may be cyclic,

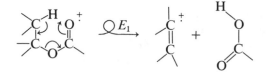

where such cyclic compounds have the same formula as the respective alkene.

Esters can eliminate neutral carboxylic acid fragments by the E_1 process, which may in many cases proceed through a six-centered mechanism:

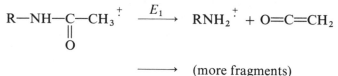

Such a case is illustrated in Figure 9.12 for *n*-propyl acetate, where the peak at m/e 42 is caused by $CH_3-CH=CH_2 \overset{+}{\cdot}$ produced by this process. Note that this elimination of esters to form alkenes is much like that of alcohols, and their spectra may not be too different. In this case, the peak at m/e 43 is more likely to be caused by $CH_3-C\equiv O^+$ (type C) than by $CH_3CH_2CH_2^+$ (type A_5), a theory that is borne out by the much lower intensity of the next peak in the homologous series at m/e 29 (43 − 14 = 29).

Amides and certain esters also fragment by the E_1 process to eliminate ketene

$$R-NH-\overset{+}{\underset{\underset{O}{\|}}{C}}-CH_3 \overset{\cdot}{} \xrightarrow{E_1} RNH_2 \overset{+}{\cdot} + O=C=CH_2$$

$$\longrightarrow \text{(more fragments)}$$

whenever an acetyl or longer group is present. This is not observed in formyl esters or amides, however.

Type *H*

This rearrangement, called the McLafferty rearrangement, is really the complement of the E_1 mechanism, since it leaves the charge on the carboxyl

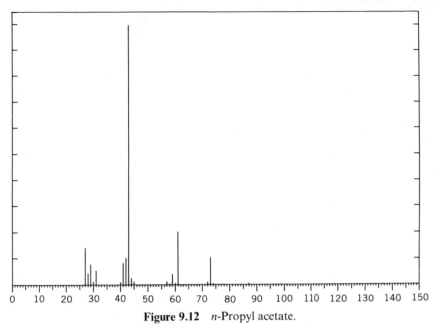

Figure 9.12   *n*-Propyl acetate.

rather than the alkene fragment:

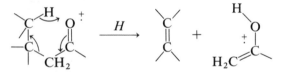

To illustrate the utility of this rearrangement in identifying fragments of similar compounds, consider the two esters in Figures 9.13 and 9.14. One is ethyl butyrate and the other butyl acetate. The butyl acetate can undergo the fragmentation

$$CH_3-CH_2-CH-CH_2^+$$
$$\quad\quad\quad H\quad O-C-CH_3$$
$$\quad\quad\quad\quad\quad\quad\quad\quad O$$

$$\xrightarrow{\;E_1\;} CH_3CH_2CH{=}CH_2^{\;+} + HO-C-CH_3$$
$$\quad\quad\quad\quad m/e\ 56 \quad\quad\quad\quad\quad\quad O$$

$$\xrightarrow{\;C\;} CH_3CH_2CH_2CH_2O\cdot + O{\equiv}\overset{+}{C}-CH_3$$
$$\quad\quad\quad\quad\quad\quad\quad\quad\quad\quad m/e\ 43$$

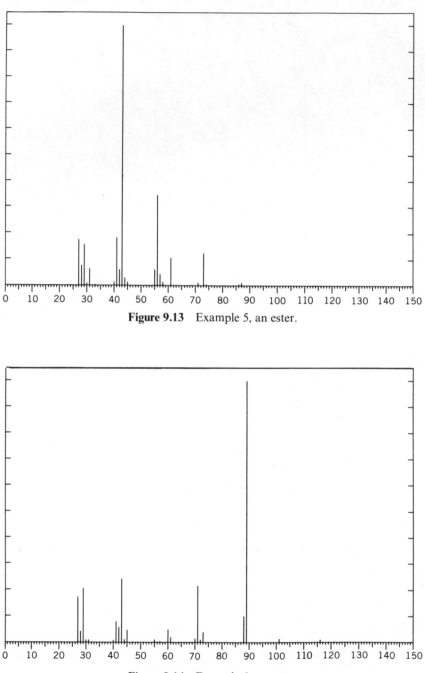

Figure 9.13 Example 5, an ester.

Figure 9.14 Example 6, an ester.

but cannot undergo the type *H* (McLafferty) rearrangement. The ethyl butyrate, on the other hand, has a three-carbon chain to the left of the carbonyl as required for the Type *H* rearrangement and undergoes the fragmentation

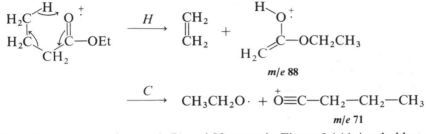

m/e 88

$$\xrightarrow{C} \quad CH_3CH_2O\cdot \; + \; \overset{+}{O}\!\!\equiv\!\!C\!\!-\!\!CH_2\!\!-\!\!CH_2\!\!-\!\!CH_3$$

m/e 71

Since the strong peaks at *m/e* 71 and 88 occur in Figure 9.14 it is ethyl butyrate, and Figure 9.13 is butyl acetate. This mechanism has been established by synthesizing

$$CD_3CH_2CH_2COOCH_2CH_3$$

and noting that the *m/e* 88 fragment increases in mass by 1 since the deuterium from the δ-carbon is incorporated. To reiterate, the E_1 cyclic elimination of esters produces a charged alkene fragment and a neutral carboxylic acid fragment, and may occur from any ester whose alcohol portion is two carbons or more. Esters can undergo Type *H* elimination when there are three or more carbons in addition to the carbonyl carbon in the acid part; that is, when there is a hydrogen γ to the carbonyl.

In a similar fashion, nitriles having γ-hydrogens will eliminate by the McLafferty rearrangement.

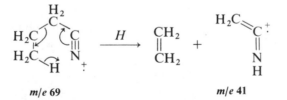

m/e 69   *m/e* 41

This rearrangement is also observed in ketones, acids, aldehydes, alkenes, alkylbenzenes, alkyl heterocycles, aryl ethers, and amides. The three atoms of the side chain, in fact, need not be carbon but can be C, N, or O, and the double bond can be a carbonyl group or part of an aromatic system.

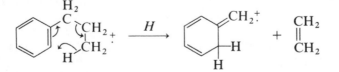

Type E_2

Closely related to this last case of Type H elimination is the E_2 cleavage

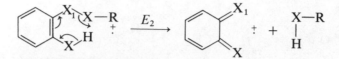

which is characteristic of *cis*-double bonds and of *ortho*-substituted aromatic systems. Here, each X may be C, N, O, or S, and X_1 may be a carbonyl. This explains why benzyl alcohol (Figure 9.15) shows almost no M-18 peak, while *o*-tolyl alcohol (Figure 9.16) shows its greatest peak at M-18.

Type F

Type F eliminations are usually two-stage processes, where a type A cleavage is further fragmented after rearrangement to produce an ion that cannot easily be explained by any simple one-step process:

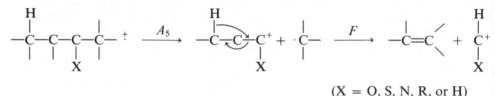

$$(X = O, S, N, R, \text{ or } H)$$

For example, the peak at m/e 31 in 4-octanol (Figure 9.17) is due to CH_2OH, which would require two bond scissions and a rearrangement. This is clearly neither likely nor energetically favorable. If, instead, we assume a type A_5 cleavage, followed by the type F rearrangement and elimination, we can explain the peak more easily:

$$CH_3-CH_2-CH_2-CH_2-\overset{+}{\underset{\underset{OH}{|}}{CH}}-C_3H_7 \xrightarrow{\;A_5\;}$$

m/e 130

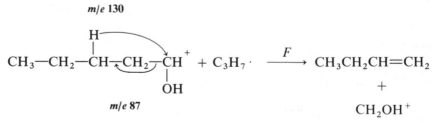

m/e 87

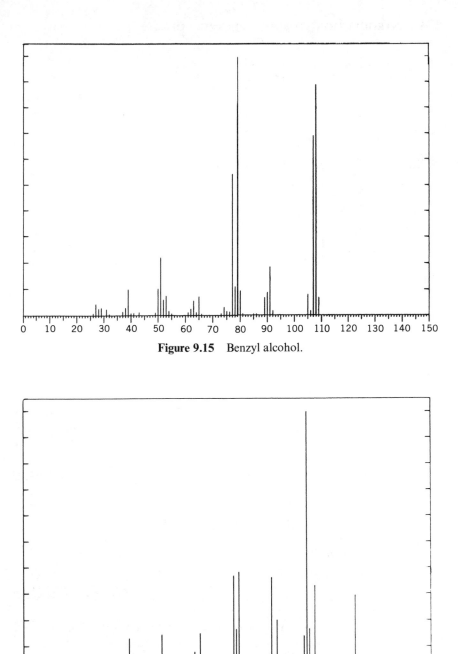

Figure 9.15   Benzyl alcohol.

Figure 9.16   *o*-Tolyl alcohol.

273

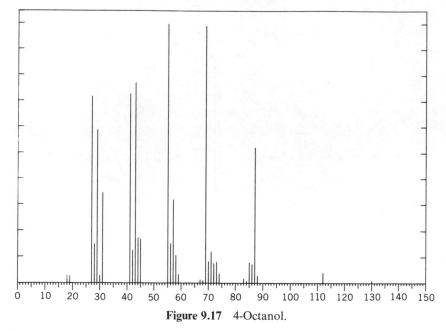

Figure 9.17 4-Octanol.

Type *G*

Type *G* elimination is found in spectra of ethers, secondary and tertiary amines, and dialkyl sulfides. It is much like type *F* in that secondary fragmentation is involved and that peaks at 30 (CH_2—NH_2^+), 31 (CH_2=OH^+), and 47 (CH_2=SH) are prominent. In this fragmentation pathway, the product of a type *B* fragmentation can rearrange to generate these small stable ions:

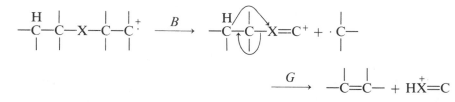

Consider the spectrum in Figure 9.18. The parent ion at m/e 73 is odd and thus contains an odd number of nitrogens. The large peak at m/e 58 suggests a labile methyl group (M-15). The peak at m/e 44 also suggests an ethyl group loss. If the compound contains one nitrogen, we are left with $73 - 14 = 59$ amu, suggesting a saturated C_4 compound. The possibilities are *n*-butyl-amine, methylpropylamine, diethylamine, *N,N*-dimethylethylamine, and

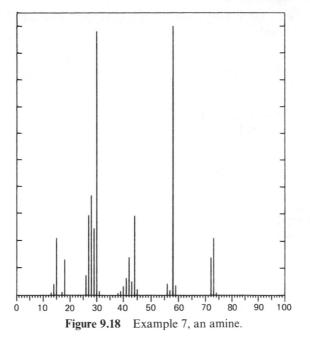

Figure 9.18   Example 7, an amine.

iso-butylamine. The major amine fragmentation process is type B,

$$R-\underset{\underset{X}{|}}{\overset{|}{C}}-R^{\overset{+}{\cdot}} \xrightarrow{\quad B \quad} R\cdot + \overset{+}{X}=\overset{|}{C}-R$$

and this leads to consideration of the fragment at m/e 58 as the most prom-ising type B fragment. The loss of the methyl group requires that an ethyl group be present so that the stabilized fragment will have one carbon left. This is only possible in diethylamine and N,N-dimethylethylamine:

$$CH_3-CH_2-NH-CH_2-CH_3^{\overset{+}{\cdot}} \xrightarrow{\quad B \quad} CH_3-CH_2-\overset{+}{N}H=CH_2 + CH_3\cdot$$
$$m/e\ \mathbf{58}$$

$$CH_3-\underset{\underset{CH_3}{|}}{N}-CH_2-CH_3^{\overset{+}{\cdot}} \xrightarrow{\quad B \quad} CH_3-\underset{\underset{CH_3}{|}}{\overset{+}{N}}=CH_2 + CH_3\cdot$$
$$m/e\ \mathbf{58}$$

Of these two structures, only one can undergo type G cleavage to explain the large peak at m/e 30:

$$\underset{\text{H}}{\overset{\text{H}}{\underset{\text{CH}_2}{\overbrace{}}}} \text{CH}_2 \underset{}{\overset{}{\frown}} \text{CH}_2 \overset{}{\frown} \text{NH}{=}\text{CH}_2{}^+ \xrightarrow{\;G\;} \text{CH}_2{=}\text{CH}_2 + \text{H}_2\overset{+}{\text{N}}{=}\text{CH}_2$$

$$\underset{m/e\ 30}{}$$

$$\underset{\underset{\text{CH}_2}{|}}{\text{CH}_3{-}\text{N}{=}\text{CH}_2{}^+} \longrightarrow \text{has no } \beta\text{-hydrogens}$$

The compound is thus diethylamine.

SUMMARY OF FRAGMENTATION PATTERNS BY FUNCTIONAL GROUP

In this section, we present a summary of the types of cleavages that various types of compounds are expected to undergo. Many of the spectra we have already seen can serve as examples of several types of fragmentation, and you should reexamine them and try to assign all of the major peaks based on the following discussion.

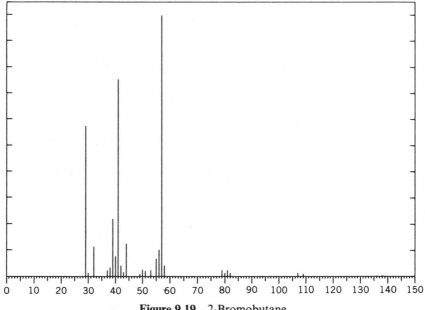

Figure 9.19 2-Bromobutane.

Alkanes

Alkanes tend to form an homologous series of peaks from M^+, going down in steps of M-14 corresponding to the loss of the next CH_2 group. Branched alkanes will show decreased intensity parent ions and loss of fragments, such as M-15 for CH_3 and M-29 for ethyl branches. The site of the branching can often be determined by examining the nature of the fragments.

Alkyl Halides

Alkyl halides undergo loss of X· to leave R^+ by type A_5 cleavage and may fragment on either side of the halogen atom by type B cleavage. The major peaks in the 2-bromobutane spectrum in Figure 9.19 are:

$$CH_3-\overset{+}{\underset{\underset{\displaystyle Br}{|}}{CH}}-CH_2-CH_3 \cdot \overset{A_5}{\longrightarrow} CH_3-\overset{+}{CH}-CH_2-CH_3$$

$$m/e\ 136 \qquad\qquad\qquad\qquad m/e\ 57$$

$$\overset{E_1}{\longrightarrow} CH_2{=}CH-CH_2-CH_3 \overset{+}{\cdot} \overset{A_3}{\longrightarrow} CH_2{=}CH-CH_2^+$$

$$m/e\ 56 \qquad\qquad\qquad m/e\ 41$$

$$\overset{B}{\longrightarrow} CH_3-CHBr^+$$

$$m/e\ 107$$

$$\overset{A_1}{\longrightarrow} CH_3-CH_2^+$$

$$m/e\ 29$$

Alkyl bromides always have pairs of peaks separated by two amu, caused by the two isotopes ^{79}Br and ^{81}Br, which have nearly equal abundance. Alkyl chlorides have pairs of peaks separated by one amu, caused by ^{35}Cl and ^{36}Cl, where the upper peak is always about 25% of the lower one. Alkyl iodides show only single peaks since there is only one isotope of iodine. Iodine can stabilize a positive charge more readily than the lower halogens and sometimes forms the $CH_2{=}I^+$ fragment by type G cleavage, observed at m/e 141. Type B cleavages of one of the other C—C bonds to the carbon bearing the iodine are favored over the lower halides. In alkyl fluorides, one of the strongest peaks is the M-1 peak, corresponding to RF^+ having lost H·.

Alcohols

Alcohols are noted for their extremely weak parent ion and strong peaks at M-18. Branched alcohols often lose methyl groups and other small side chains easily, and when several types of fragmentation can occur, the charge usually stays with the larger fragment. Fragmentation of alcohols by type B cleavage is most common,

$$R-CHOH-R \overset{+}{\cdot} \xrightarrow{\;B\;} RCH=\overset{+}{O}H + R\cdot$$

and primary alcohols frequently show peaks at m/e 31 for this reason:

$$CH_3(CH_2)_4CH_2OH\overset{+}{\cdot} \xrightarrow{\;B\;} CH_3(CH_2)_4\cdot + CH_2=\overset{+}{O}H$$
$$m/e\ 31$$

This is apparent in the spectrum of hexanol in Figure 9.5.

It has been shown by deuterium labeling experiments that the E_1 elimination of water does not generally involve the α-hydrogens, but rather those from more distant carbons. This can also account for the common elimination of ethylene from the resulting M-18 fragment, leading to an M-46 peak of substantial intensity. This is, in fact, the largest peak in the spectrum of 1-hexanol, and may be formed by a six-centered mechanism in a concerted reaction involving the elimination of both water and ethylene:

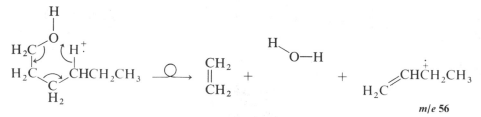

$$m/e\ 56$$

Alkenes

The alkenes undergo fragmentation primarily by the A_3, E_2, and H routes. Cyclic alkenes also undergo retro-Diels–Alder cleavage (type D). Unfortunately, most alkene spectra represent a homologous series of loss of each individual carbon, and are not easily distinguishable on the basis of mass spectroscopy alone, because double bonds tend to migrate and *cis–trans* isomers to interconvert during fragmentation.

For example, the spectrum of 2-methyl-2-hexene in Figure 9.20 shows a strong M-29 peak at m/e 69 corresponding to the A_3 cleavage

$$CH_3-CH_2-CH_2-CH=\overset{\underset{\textstyle |}{CH_3}}{C}-CH_3 \xrightarrow{\;A_3\;}$$

$$CH_3-CH_2\cdot + \overset{+}{C}H_2-CH=\overset{\underset{\textstyle |}{CH_3}}{C}-CH_3$$
$$m/e\ 69$$

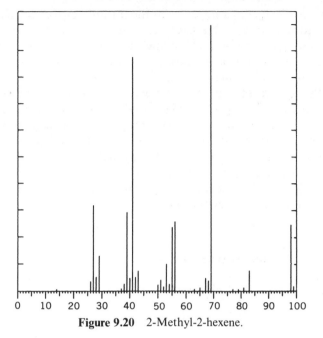

Figure 9.20 2-Methyl-2-hexene.

but also shows peaks corresponding to C_4H_8 (m/e 56), C_3H_5 (m/e 41), and C_3H_3 (m/e 39), which cannot be explained by any simple fragmentation without double bond migration.

$$
\underset{m/e\ 56}{\overset{\underset{|}{CH_3}}{CH_2{=}C{-}CH_3}}
\qquad
\underset{m/e\ 41}{CH_2{=}CH{-}CH_2^+}
\qquad
\underset{m/e\ 39}{H{-}C{\equiv}C{-}\overset{+}{C}H_2}
$$

"Random" Rearrangements

There are some rearrangements that are not predictable or expected from simple organic chemistry or the fragmentation theories we have discussed in this chapter. These are the so-called "random" rearrangements and are found most commonly in the mass spectra of alkanes, alkenes, and some aromatic hydrocarbons.

For example, in the spectrum of 2,2-dimethylbutane (neohexane) a strong peak is observed at m/e 43,

$$
\underset{\underset{|}{CH_3}}{\overset{\overset{|}{CH_3}}{CH_3{-}C{-}CH_2CH_3}}
\longrightarrow
\underset{m/e\ 43}{CH_3\overset{+}{C}HCH_3}
$$

which can only be due to a fragment of the formula $C_3H_7^+$. Such a fragment cannot be formed in any simple rearrangement process and must involve breaking two carbon bonds and the migration of a hydrogen atom.

Unsaturated hydrocarbons show so many such rearrangements that it is difficult to tell the spectra of isomers apart. For example, Beynon[6] has pointed out that the spectra of 2-methyl-2-butene, 3-methyl-1-butene, 2-methyl-1-butene, 2-pentene, and 1-pentene are virtually identical, despite the fact that we might have expected rather different fragmentation patterns, *a priori*.

Similarly, the mass spectrum of 1,1,1-trideuteroethane (CD_3—CH_3) shows ions that can be ascribed to CHD_2^+ as well as to CD_3^+, and ions due to methyl group loss from dideuterated xylenes (CH_2D—C_6H_4—CH_2D) occur at *m/e* 91, 92, and 93, indicating that hydrogen/deuterium migration must have occurred before fragmentation.

Ethers

Ethers undergo primarily type A_5 cleavage,

$$CH_3-CH_2-CH_2-\overset{+\cdot}{O}-CH_2-CH_2-CH_3 \quad \xrightarrow{A_5}$$

$$CH_3-CH_2-CH_2-\dot{O} + \overset{+}{C}H_2-CH_2-CH_2$$
$$m/e\ 43$$

and type G cleavage,

$$\overset{H}{\longrightarrow}\ CH_3-\overset{H}{C}H-CH_2-O{=}CH_2^+ + \dot{C}H_2-CH_3$$

$$\xrightarrow{G}\ CH_3-CH{=}CH_2 + H\overset{+}{O}{=}CH_2$$
$$m/e\ 31$$

both of which are seen in Figure 9.8. Both fragments of an asymmetrical ether will be observed for type A_5 cleavage, and ethers branched α to the oxygen will show peaks at higher amu by type G cleavage.

Aldehydes

Aldehydes undergo type C cleavage to form *m/e* 29 and M-1 peaks:

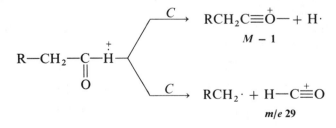

$$\xrightarrow{C}\ RCH_2C{\equiv}\overset{+}{O}- + H\cdot$$
$$M - 1$$

$$R-CH_2-\overset{+}{\underset{\parallel}{C}}-\overset{+}{H}$$
$$O$$

$$\xrightarrow{C}\ RCH_2\cdot + H-C{\equiv}\overset{+}{O}$$
$$m/e\ 29$$

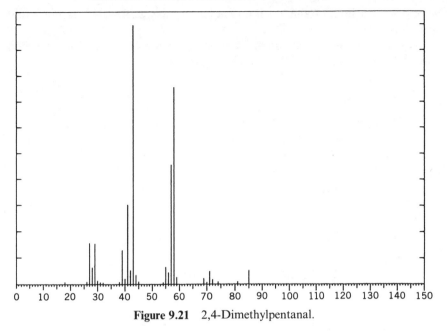

Figure 9.21   2,4-Dimethylpentanal.

They also undergo type *H* rearrangements to lose alkene fragments when γ-hydrogens are present. Figure 9.21 illustrates this point for 2,4-dimethyl-pentanal, where a substantial peak occurs at *m/e* 58 corresponding to iso-butene formed by the type *H* process. Note that this is the only large *even* mass fragment and thus the only product caused by a rearrangment and loss of a neutral molecule.

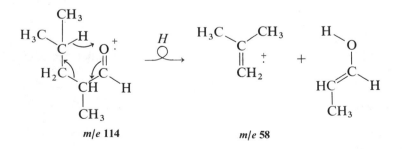

Ketones

Ketones undergo the same type *C* cleavages and type *H* rearrangements as do aldehydes. Asymmetric ketones undergo two types of type *C* cleavage, with loss of the larger fragment as an uncharged species favored, as illustrated in Figure 9.10.

Ketones that have 3-carbon or longer chains on both sides of the carbonyl can undergo double McLafferty rearrangements.

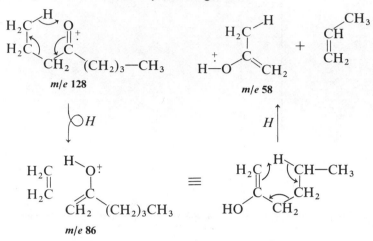

Esters and Carboxylic Acids

Carboxylic acids of larger molecules are not very volatile and are less used in mass spectroscopy than their corresponding esters; their spectra are characterized by loss of OH (M-17) and loss of COOH (M-45). Esters can undergo type C cleavage,

$$CH_3CH_2CH_2-\underset{\underset{O}{\|}}{C}-OCH_2CH_3^{\ddagger} \xrightarrow{\;\;C\;\;}$$

$$CH_3CH_2CH_2-C{\equiv}\overset{+}{O} + \overset{\cdot}{O}CH_2CH_3$$
$$m/e\ 73$$

and type H cleavage when a γ-hydrogen is present. This is illustrated in Figure 9.14 for ethyl butyrate and is discussed in detail under type H cleavage (p. 268).

Amines

The primary fragmentation processes in amines are types B and G. In primary amines this always leads to a peak at m/e 30:

$$RCH_2\overset{+}{NH_2}{\overset{\cdot}{}}\ \xrightarrow{\;\;B\;\;}\ R\cdot\ +\ CH_2{=}NH_2^+$$
$$m/e\ 30$$

In secondary amines the peak will be at $R_1 + 29$ and in tertiary amines at $R_1 + 28$ and $R_2 + 28$:

$$RCH_2-NHR_1^+ \xrightarrow{\;B\;} R\cdot + CH_2=\overset{+}{N}HR_1 \qquad \text{and so forth.}$$

In all cases, the largest R group has the greatest probability of being lost as a radical. Secondary amines can also form the peak at m/e 30 by a type B cleavage followed by a type G rearrangement:

$$RCH_2-CH_2-NH-CH_2R' \xrightarrow{\;B\;}$$

$$RCH\overset{H}{\frown}CH_2\overset{\frown}{-}NH=CH_2^+ + R\cdot \xrightarrow{\;G\;} RCH=CH_2 + HNH=CH_2^+$$
$$m/e\ 30$$

Secondary and tertiary amines having branched α-carbons can form larger fragments by the type B cleavage followed by the G process. Figure 9.22 shows the spectrum of di-isopropylamine, which shows the following fragmentation patterns:

$$(CH_3)_2CH-NH-CH(CH_3)_2 \xrightarrow{\;B\;} (CH_3)_2CH-\overset{+}{N}H=CHCH_3$$
$$m/e\ 101 \qquad\qquad m/e\ 86$$

$$\xrightarrow{\;G\;} CH_2=CH + H\overset{+}{N}H=CH-CH_3$$
$$\overset{\displaystyle CH_3}{|}$$
$$m/e\ 44$$

Amines also lose fragments by N—C bond cleavage, but to a lesser degree:

$$R-NH-R \xrightarrow{\;A_5\;} R-NH^+ + R\cdot$$

Note that such fragments in single nitrogen compounds will be at *even* masses, while rearrangement products will be at *odd* masses, in contrast to non-nitrogen-containing compounds where the reverse holds true. In the spectrum of di-isopropylamine, a small peak at m/e 58 is noted from this fragmentation.

$$CH_3\overset{\displaystyle CH_3}{\underset{|}{C}}H-\overset{+}{N}H + \overset{\cdot}{C}H(CH_3)_2$$
$$m/e\ 58$$

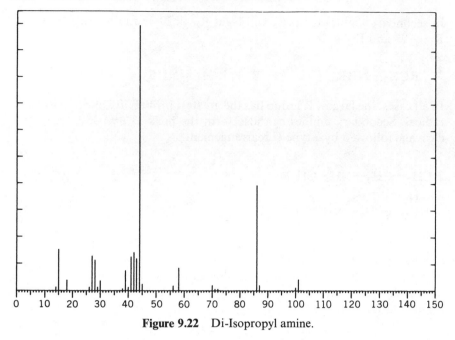

Figure 9.22   Di-Isopropyl amine.

Aromatic Compounds

Substituted benzenes fragment according to cleavage type A_4 to form the tropyllium ion (m/e 91). This usually fragments further to lose acetylene and form the cyclopentadienyl cation (m/e 65). Alkyl side chains having γ-hydrogens rearrange by mechanism H to lose ethylene. Both of these fragments are observed in the spectrum of benzyl methyl ether, Figure 9.23.

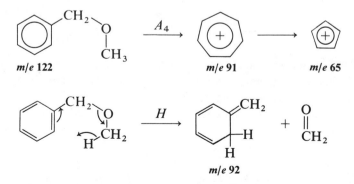

Similar patterns are observed for alkyl naphthalenes.

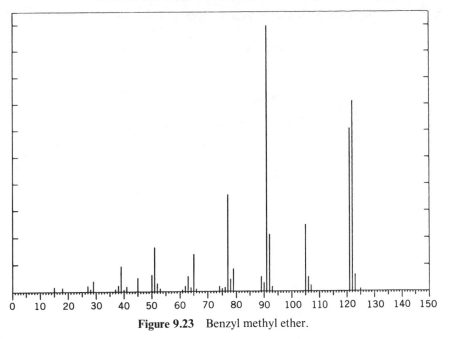

Figure 9.23   Benzyl methyl ether.

Alkyl benzyl amines, on the other hand, do not show E_2 cleavage because the nitrogen atom stabilizes the positive charge more effectively.

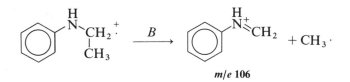

Ortho-substituted aromatic compounds undergo E_2 rearrangement as was illustrated in Figure 9.16 for *o*-tolyl alcohol. Similarly, *o*-methyl toluate shows this rearrangement:

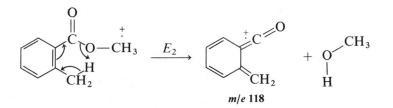

Phenols show strong M-28 peaks by loss of CO and M-29 due to loss of HCO.

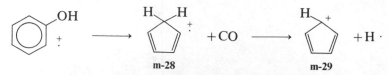

Benzyl alcohols show strong M-1 peaks due to loss of H and formation of hydroxytropyllium ions, (as well as peaks at M-29 caused by loss of CO from the hydroxytropyllium ion.) Nitrobenzenes show loss of NO and NO_2 as well as the usual fragmentation of the benzene ring.

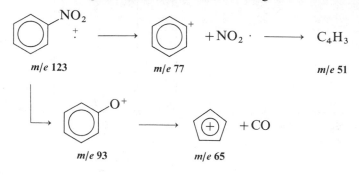

FRAGMENTATION PATTERN SUMMARY

Simple Cleavage

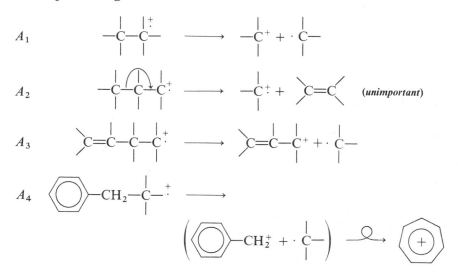

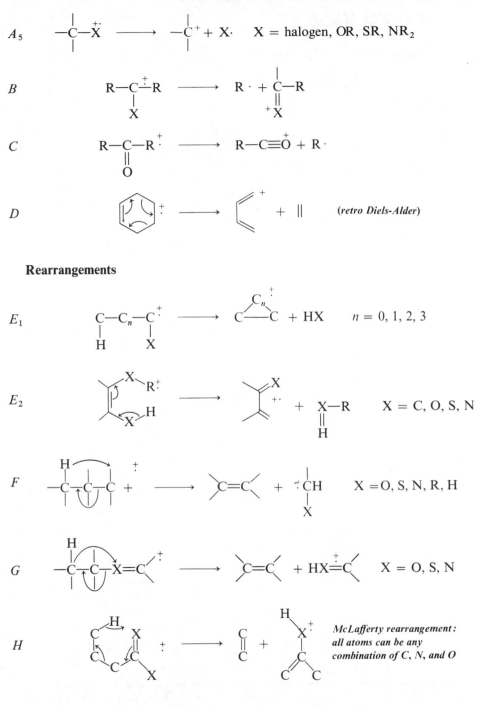

A_5

$$-\overset{|}{\underset{|}{C}}-\overset{+\cdot}{X} \longrightarrow -\overset{|}{\underset{|}{C}}{}^{+} + X\cdot \qquad X = \text{halogen, OR, SR, NR}_2$$

B

$$R-\overset{+}{\underset{X}{C}}-R \longrightarrow R\cdot + \overset{|}{\underset{\overset{||}{{}^{+}X}}{C}}-R$$

C

$$R-\overset{+\cdot}{\underset{\overset{||}{O}}{C}}-R \longrightarrow R-C\equiv\overset{+}{O} + R\cdot$$

D

$$\longrightarrow \qquad + \, || \qquad (\textit{retro Diels-Alder})$$

Rearrangements

E_1

$$\overset{}{\underset{H}{C}}-C_n-\overset{+\cdot}{\underset{X}{C}} \longrightarrow C{-\!-\!}C + HX \qquad n = 0, 1, 2, 3$$

E_2

$$\longrightarrow \qquad + \quad \underset{\overset{||}{H}}{X-R} \qquad X = C, O, S, N$$

F

$$-\overset{H}{\underset{|}{C}}-\overset{|}{\underset{|}{C}}-\overset{|}{\underset{|}{C}} + \longrightarrow \;\; C{=}C \;\; + \;\; \overset{\cdot}{\underset{X}{C}}H \qquad X = O, S, N, R, H$$

G

$$-\overset{H}{\underset{|}{C}}-\overset{|}{\underset{|}{C}}-X{=}C \longrightarrow C{=}C \;\; + \;\; HX{\overset{+\cdot}{=}}C \qquad X = O, S, N$$

H

$$\longrightarrow \quad \overset{C}{\underset{C}{||}} \;\; + \qquad \begin{array}{l}\textit{McLafferty rearrangement:}\\ \textit{all atoms can be any}\\ \textit{combination of C, N, and O}\end{array}$$

PROBLEMS

Problem 9.1

An acrid-smelling liquid,
MW = 60.0211

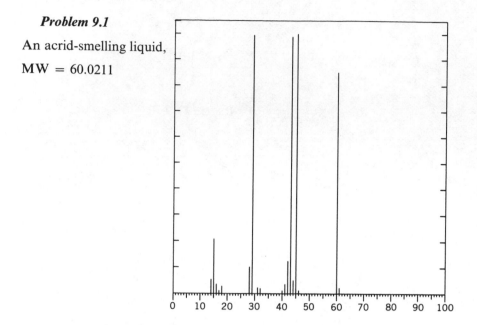

Problem 9.2

An oxygen-containing compound, MW = 100.0888

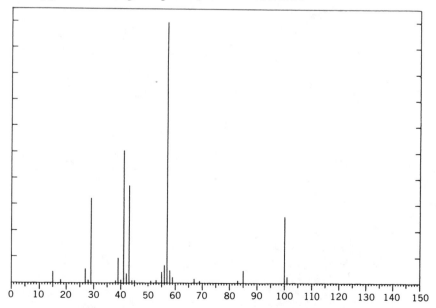

Problem 9.3

A hydrocarbon, MW = 120.0939

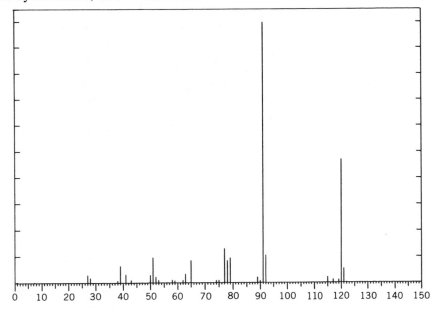

Problem 9.4

A fruity-smelling liquid, MW = 88.0524

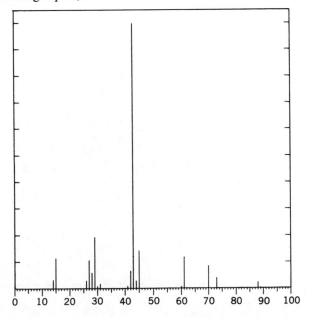

Problem 9.5

A hydrocarbon, MW = 96.09396

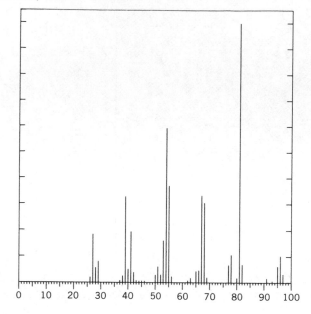

Problem 9.6

A strong-smelling liquid, MW = 123.9524

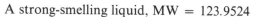
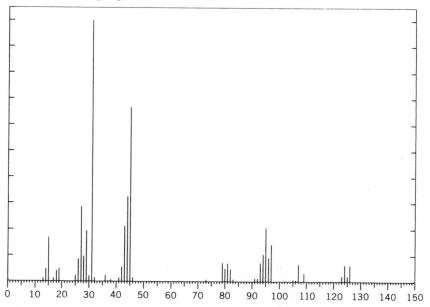

Problem 9.7

MW = 123.0320

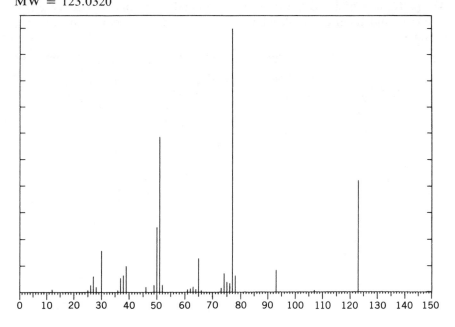

Problem 9.8

MW = 150.0681

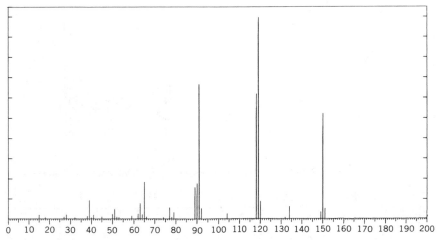

Problem 9.9

IR, pmr, mass, and ^{13}C spectra (decoupled and off-resonance decoupled) of a compound having a molecular weight of 151.9837. No significant UV absorption.

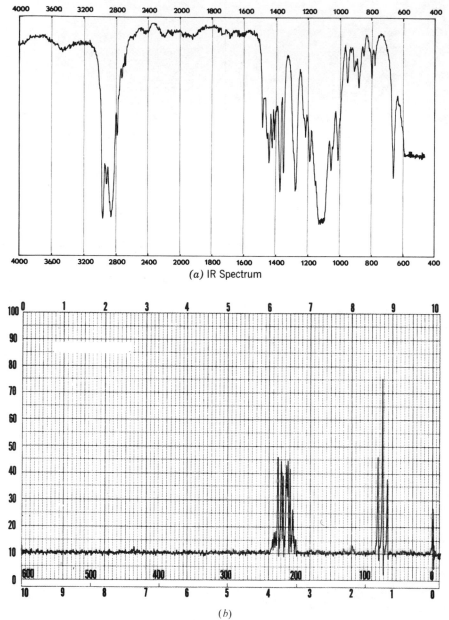

(a) IR Spectrum

(b)

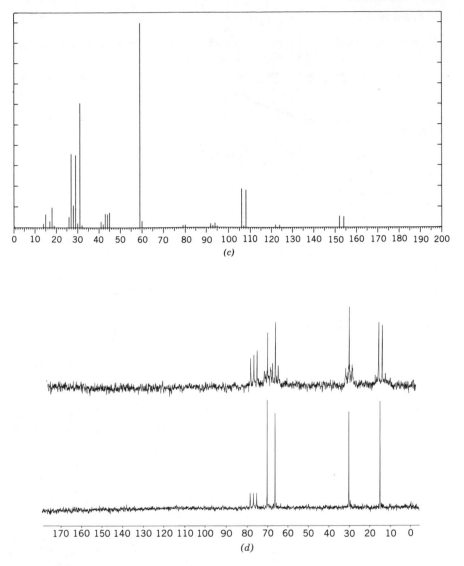

(c)

(d)

Problem 9.10

IR spectrum (Nujol mull), pmr, mass and ^{13}C spectra of a compound having a molecular weight of 135.0685. UV absorption observed.

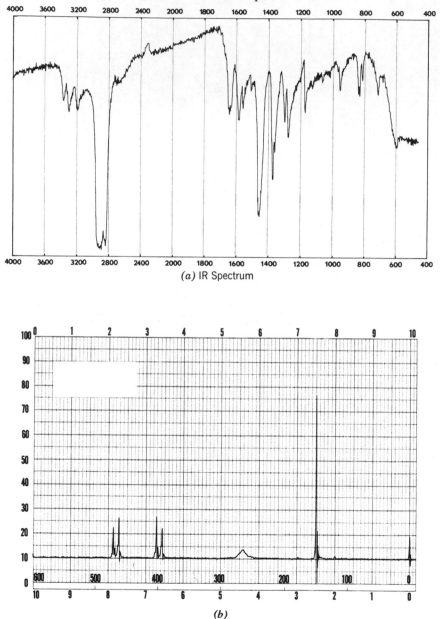

(*a*) IR Spectrum

(*b*)

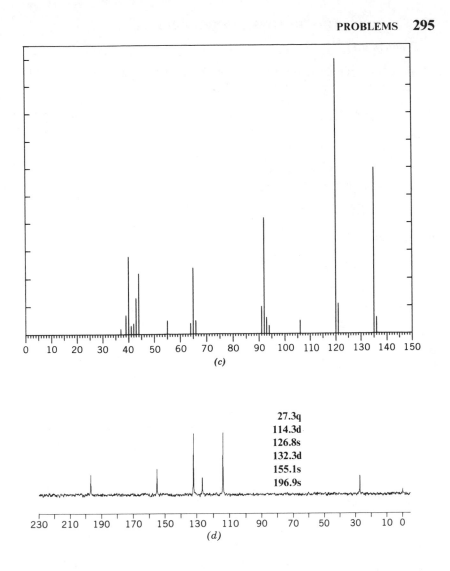

(c)

27.3q
114.3d
126.8s
132.3d
155.1s
196.9s

(d)

Problem 9.11

IR, pmr, mass, and ^{13}C spectra of a compound having a molecular weight of 132.0575. UV $\lambda_{max} = 248$ nm, $\varepsilon = 15{,}000$.

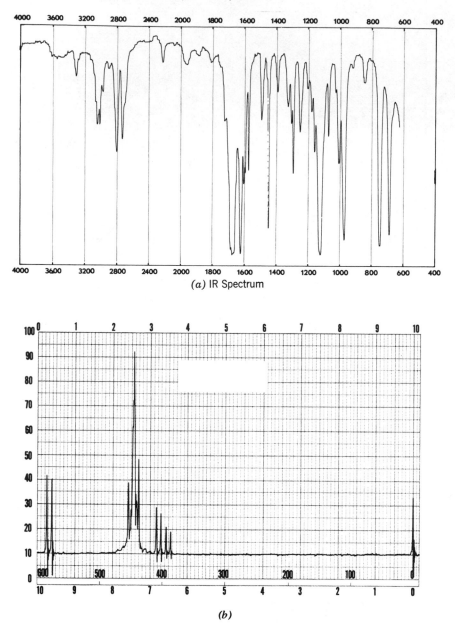

(a) IR Spectrum

(b)

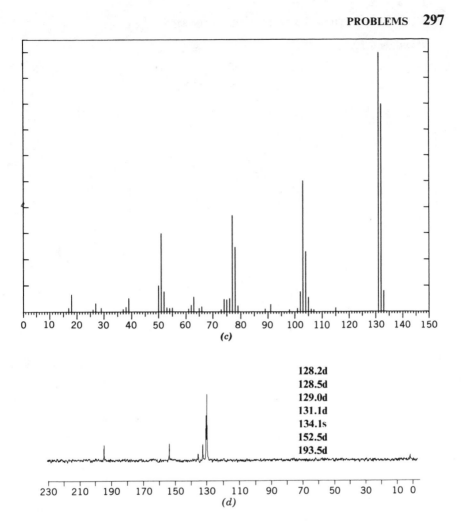

(c)

128.2d
128.5d
129.0d
131.1d
134.1s
152.5d
193.5d

(d)

Problem 9.12

IR, pmr, mass, and ^{13}C spectra of a compound having a strong mint odor.
MW = 152.0473. UV absorptions observed.

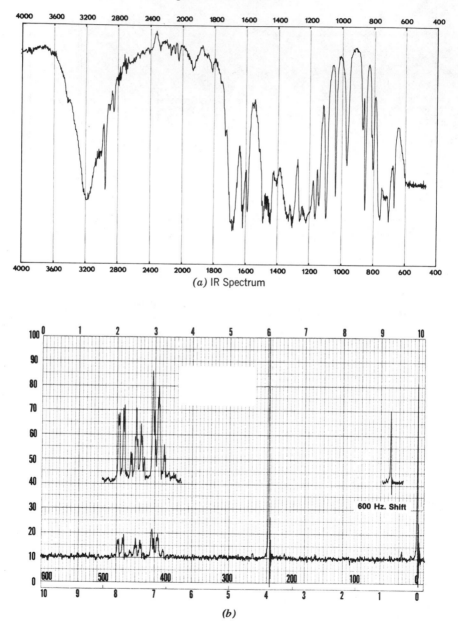

(a) IR Spectrum

(b)

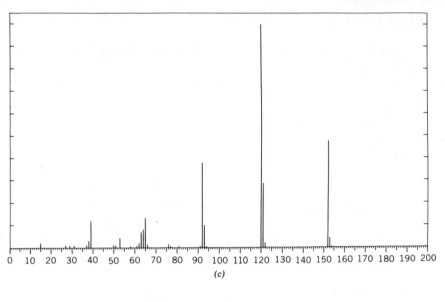

(c)

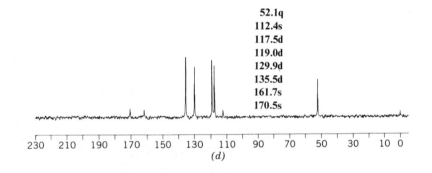

52.1q
112.4s
117.5d
119.0d
129.9d
135.5d
161.7s
170.5s

(d)

REFERENCES

1. K. Biemann, *Mass Spectrometry, Organic Chemical Applications*, McGraw-Hill, New York, 1962.
2. F. W. McLafferty, *Interpretation of Mass Spectra—An Introduction*, W. A. Benjamin, New York, 1967.
3. H. Budzikiewicz, C. Djerassi, and D. Williams, *Interpretation of Mass Spectra of Organic Compounds*, Holden-Day, San Francisco, 1964.
4. M. Hamming and N. Foster, *Interpretation of Mass Spectra of Organic Compounds*, Academic Press, New York, 1972.
5. G. W. A. Milne, Ed., *Mass Spectrometry: Techniques and Applications*, Wiley-Interscience, New York, 1971.
6. J. H. Beynon, *Mass Spectrometry and Its Applications to Organic Chemistry*, Elsevier, Amsterdam, 1960.

COLLECTIONS OF SPECTRA

Heller, S. R., and G. W. A. Milne, "EPA/NIH Mass Spectral Data Base," Vols. 1–4, U.S. Government Printing Office, NSRDS–NSB–63, Washington, D.C., 1978.

Stenhagen, E., S. Abrahamson, and F. W. McLafferty, *Atlas of Mass Spectral Data*, Wiley-Interscience, New York, 1969.

CHAPTER TEN

RAMAN SPECTROSCOPY

While Raman spectroscopy has been commonly used by the spectroscopist and inorganic chemist for some years, it is now increasingly used by the organic chemist as an additional tool for structure determination. The Raman spectrum provides complementary information to the IR spectrum, in that both are vibrational spectra. Often, the weakest lines in the IR spectrum will prove to be the strongest lines in the Raman spectrum. To understand why this is so, we need to understand the Raman effect.

THE RAMAN EFFECT

To discuss Raman scattering, it is convenient to consider the particle model of light, which is based on small packets of energy called *photons*. If a photon interacts with a molecule and is scattered in an elastic collision, we might expect to observe some light scattered perpendicular to the light source. If this light has the same frequency as the incident light because of an elastic collision, we call this *Rayleigh scattering*. If, on the other hand, the light is of a different frequency than the incident light because of an inelastic collision and energy transfer between a photon and a sample molecule, we call this *Raman scattering*.

Both scattering effects are fairly weak. The light observed due to Rayleigh scattering is usually about 10^{-3} the intensity of the incident light and that due to Raman scattering is about 10^{-6} or 10^{-7} the intensity of the incident light.

When a single coherent frequency laser source is used in Raman studies, there are two types of Raman scattering possible: those lines whose frequencies are below and those whose frequencies are above that of the laser. The lines having lower frequencies are more common and are called the *Stokes lines*. These arise from collisions between photons and molecules in the ground state, resulting in a less energetic photon (lower frequency) and an excited molecule. The lines that have frequencies higher than the laser are called *anti-Stokes lines*. These arise when a photon collides with a molecule in an excited state and the energy is transferred to the photon as the

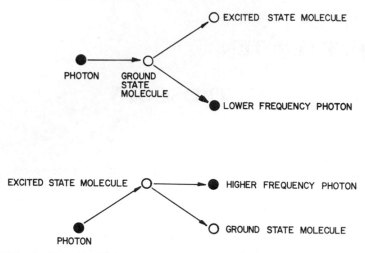

Figure 10.1 Stokes and anti-Stokes transitions. In the upper drawing, a photon collides with a ground-state molecule, producing an excited-state molecule and a lower energy photon, causing a Stokes transition. In the lower drawing, a photon collides with an excited-state molecule transferring energy to the photon, causing a higher frequency photon and a ground-state molecule. This is an anti-Stokes line.

molecule relaxes to its ground state. This is illustrated in Figure 10.1. Stokes lines are much stronger than anti-Stokes lines since the ground state population is much greater than the excited state population; most Raman studies have been of the Stokes lines. We will consider only the Stokes lines in this chapter.

POLARIZABILITY: THE DIFFERENCE BETWEEN IR AND RAMAN LINES

In order for a molecule to absorb in the IR, there must be a change in the dipole moment of the molecule with the vibration. Thus we write the carbonyl stretching vibration, for example, as

$$\underset{/}{\overset{\backslash}{C}}=O \quad \xrightarrow{h\nu} \quad \underset{/}{\overset{\backslash}{\overset{+}{C}}}-\overset{-}{O}$$

to show a clear change in the dipole moment during the vibration. By contrast, in Raman spectroscopy the vibrations involved are generally much more symmetric and result in less overall change in the dipole moment. Instead, the selection criterion for Raman absorption is a change in the molecular *polarizability*.

This polarizability can be regarded as the ability of the molecule to be polarized or to have a dipole *induced* in it, and results from changes in the distribution of the electron cloud during the stretching vibration. If the dipole can be induced more easily during this vibration, then we say that there has been a *change* in the *polarizability* of the molecule.

The electrical field in question in Raman spectroscopy is, of course, that provided by the electrical component of the light itself. It is usually safe to say that compounds showing the greatest change in polarizability show the least change in dipole moment, and thus the lines that are weakest in the IR spectrum may be stronger in the Raman spectrum.

A simple pictorial model of two of the stretching modes of CO_2 serves to illustrate the difference between these two concepts. In Figure 10.2*a* we see the symmetrical stretch of CO_2 in which both oxygens move farther away from the carbon at the same time. There is no change in the dipole moment for this stretching mode and it is not observed in the IR spectrum. However, the stretching of the molecule in this way perturbs the electron distribution and causes a change in the polarizability in the stretched state from that in the unstretched state, thus leading to a line in the Raman spectrum.

By contrast, a stretching mode such as that shown in Figure 10.2*b* shows a change in the dipole moment, since the oxygens move asymmetrically with respect to the carbon. This mode shows a line in the IR spectrum. On the other hand, there is no change in the polarizability of the molecule for this vibration since any change in the electron distribution caused by the move-ment of one oxygen is complemented by a similar change in the other oxygen at the other extreme of the stretch, and this mode shows no line in the Raman spectrum.

For real organic molecules, this polarizability will differ in each of the three coordinate directions, since most organic molecules are not this sym-metrical. The actual polarizability is then described by the interaction of the electric field vector with that of the molecule, and leads to the 3 × 3 matrix

$$O \leftarrow C \rightarrow O \quad \xrightarrow{h\nu} \quad O - C - O$$

(a) **symmetric stretch Raman active**

$$O \rightarrow C \rightarrow O \quad \xrightarrow{h\nu} \quad OC - O \quad \text{and} \quad O - CO$$

(b) **asymmetric stretch IR active**

Figure 10.2 The symmetric and asymmetric stretching modes of CO_2. The symmetric stretch produces no change in dipole moment, but a change in polarizability and is thus Raman active. The asymmetric stretch produces a dipole moment change, but no change in polarizability and is thus active in the IR.

called the polarizability *tensor*. Such tensors are not predictable *a priori* for large asymmetric molecules, and we will instead summarize group frequencies and intensities much as we did for IR.

MUTUAL EXCLUSION RULE

To illustrate the difference between Raman and IR absorptions, we point out that in molecules that have a center of symmetry, there will be no correspondence between the IR spectrum and the Raman spectrum, since any line that is active in the IR will be forbidden in the Raman and vice versa. A molecule having a center of symmetry is one in which a line drawn through any atom and through the center of the molecule will encounter an equivalent atom on the other side of the molecule. Simple examples of such molecules include benzene and *trans*-1,2-dichloroethylene:

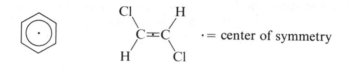

· = center of symmetry

DEPOLARIZATION RATIOS

While in other optical spectroscopic methods we report only the frequency of the absorption and the intensity, in Raman spectroscopy we also report the *depolarization ratio*. This number is the ratio of the intensity of polarized light perpendicular to the plane of the incident light over that intensity parallel to the incident light plane

$$\rho = \frac{I_\perp}{I_\parallel}$$

and is usually symbolized by ρ.

Experimentally, we place a polarizing disk which can be rotated in front of the analyzer. There is no need to place one in the incident light path when laser sources are used, since the radiation is coherent and plane polarized already. We measure the entire spectrum twice: once with the polarizer oriented parallel to the plane of the incident light and once with the polarizer oriented perpendicular to this plane. The depolarization ratio for any line is the ratio of the peak heights of the second scan divided by the first.

We find that the depolarization ratio for totally symmetric molecules is zero: no light is depolarized. For less symmetric molecules the ratio approaches the theoretical maximum of 0.75.

SPECTROMETER CONFIGURATION AND SAMPLE HANDLING

Modern Raman spectrometers operate with a monochromatic coherent laser light source. Several laser frequencies, as well as tunable lasers, are in use. The spectrum is reported as the wave number distance from the incident light frequency and will be the same regardless of the frequency of the source used, barring interactions with strong light absorption of the incident light frequency.

Sample handling is also much simpler in Raman spectroscopy than in IR spectroscopy. The sample can be placed in any of a large number of solvents, since their Raman absorptions are much more limited than in IR. In fact, solid samples can be examined quite conveniently as solid powders or even single crystals. Liquids can be placed in small quartz or Pyrex glass capillaries, where the scattered radiation is observed through the open end of the capillary. Samples can even be examined without removing them from the bottles!

In most cases, if several spectroscopic techniques are to be performed on a small amount of sample, the Raman spectrum should be obtained first, since it is totally nondestructive and does not require dilution in any solvent as does nmr or UV.

FLUORESCENCE

Raman spectra are occasionally plagued by the problem of fluorescence, the familiar effect of light emission from a compound after excitation with a certain frequency of light. This effect differs only in degree from Raman scattering, and if the compound itself fluoresces, it may be possible to eliminate the problem by changing the frequency of the exciting laser, since fluorescence usually occurs with a specific excitation frequency. More common, however, is the presence of small fluorescing impurities that confuse and swamp the actual Raman spectrum. These impurities can sometimes be doped out of the sample by addition of 1 % nitrobenzene to liquids and 1 % p-dinitrobenzene to solids. Fluorescent " background " can also be removed digitally if the Raman spectrum is acquired by a computer system.

GROUP FREQUENCIES IN RAMAN SPECTRA

Amines

The IR spectra of primary amines are occasionally obscured by intramolecular hydrogen bonding or OH groups, so that the 1° amine doublet at 3300 cm^{-1} is not visible. This is never the case in the Raman spectra, so that this technique can be used for identification of substituted amines.

Alkynes

One of the most difficult lines to see in IR spectra is the C≡C stretch of interior alkynes. Terminal alkynes give a C—H stretch at 3270–3315 cm^{-1}, but nonterminal alkynes show only a weak IR peak at 2200–2260 cm^{-1}. As the interior alkynes become more and more symmetrical, this line becomes much stronger in the Raman spectrum and represents an important identification technique for these compounds. The C≡C stretch in 3-hexyne would be totally invisible in the IR as shown in Figure 10.3a, but is shown as a strong absorption at 2233 cm^{-1} in the Raman spectrum in Figure 10.3b.

Nitriles

Some C≡N stretching lines become extremely weak in the IR when the α-carbon bears chlorine or oxygen, but are quite strong in the Raman in the range of 2240–2260 cm^{-1}. This represents an important confirming technique for nitriles, along with ^{13}C nmr.

Alkenes

The C=C stretching frequencies of alkenes are always stronger in the Raman than in the IR spectrum. This is particularly clear in the spectrum of 1-octene shown in Figure 10.4b where the C=C stretch at 1645 cm^{-1} is most prominent. Furthermore, the depolarization ratios can help prove the substitution of the double bond, as shown in Table 10.1 below.

Table 10.1   IR and Raman Characteristics of C=C Absorptions

	IR intensity	Raman ρ
RCH=CH$_2$	med	0.04
R$_2$C=CH$_2$		0.04
cis-RCH=CHR		0.05
trans-RCH=CHR		0.08
tri- and tetra-	weak	0.1

Carbonyl Compounds

While IR remains the primary method of identifying carbonyl compounds, the depolarization ratios of esters and carboxylic acids serve to set them off from other compounds in which other relevant information may be obscured (Table 10.2).

Table 10.2 IR and Raman Characteristics of Carbonyl Compounds

	IR	Raman ρ for C=O
Ketones	methyl ketones 1350–1370	0.22–0.27 acyclic, 0.28 cyclic
Aldehydes	C—H 2695–2900	0.24–0.28
Esters	1000–1250	0.1
Carboxylic Acids	OH at 3000	0.05

Cyclic Alkanes

Most cyclohexane compounds that are not fused to other rings show a strong ring breathing frequency in the Raman at 700–800 cm^{-1}. This is illustrated in Figure 10.5 for cyclohexane.

Aromatic Compounds

The following lines may be used to identify some of the common benzene substitution patterns:

	Raman line (cm^{-1})	ρ
ortho	1020–1050	0.03
	640–760	
meta	990–101	0.1
para	625–645	0.2
mono	899–1006 vs	
	1021–1035 m	
	605–625 w	

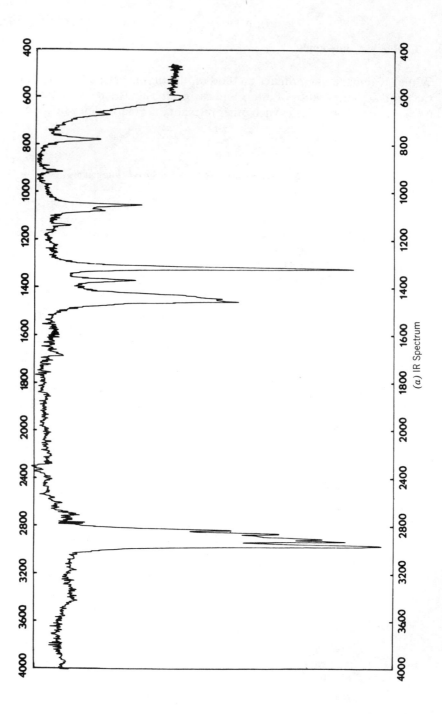

(a) IR Spectrum

308

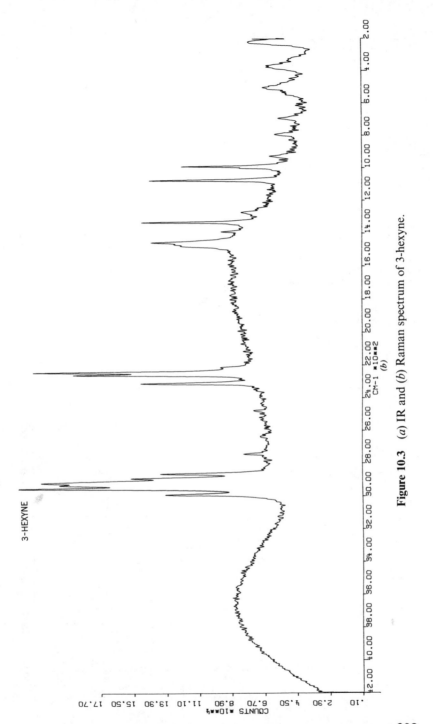

Figure 10.3 (*a*) IR and (*b*) Raman spectrum of 3-hexyne.

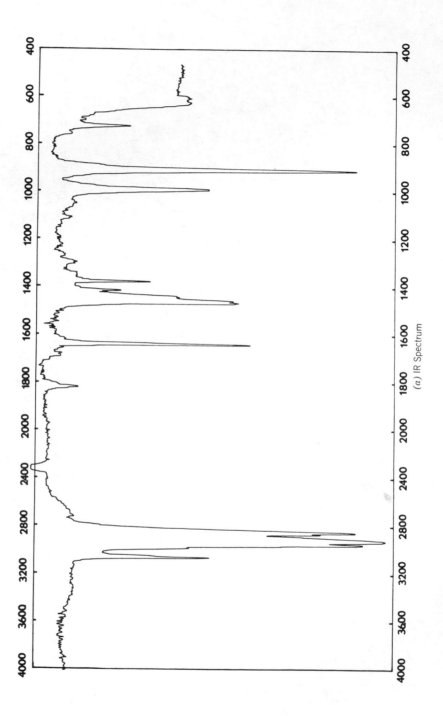

(a) IR Spectrum

310

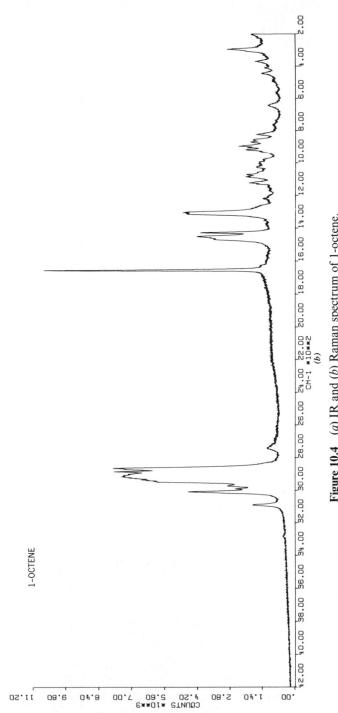

1-OCTENE

Figure 10.4 (*a*) IR and (*b*) Raman spectrum of 1-octene.

(*b*)

CM-1 *10**2

COUNTS *10**3

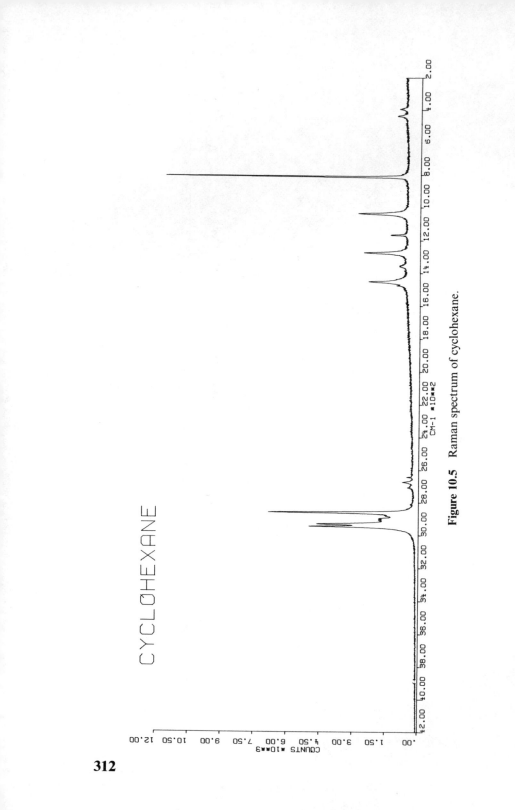

Figure 10.5 Raman spectrum of cyclohexane.

SUMMARY

The Raman technique has now become important in organic chemistry as well as in inorganic chemistry. Sample handling, even of aqueous solutions, is more versatile, making many compounds more accessible to spectroscopic analysis. Depolarization ratio measurements can often help distinguish between compounds that seem to be spectroscopically similar.

RECOMMENDED READING

Freeman, S. K., *Applications of Laser Raman Spectroscopy*, Wiley-Interscience, New York, 1974.

ANSWERS TO PROBLEMS

Problem 2.1 (C_6H_{14})

From the formula alone, it can be seen that this is obviously a simple hydro-carbon. The lack of a *gem*-dimethyl doublet around 1380 cm^{-1} clearly indicates that branching is not at the second carbon of the chain. The ethyl group peak near 790 cm^{-1} suggests there is branching at an interior carbon. The compound is therefore 3-methylpentane.

Problem 2.2 (C_8H_{18})

Neither the formula nor the bands near 3000 cm^{-1} indicate unsaturation. The compound is a saturated hydrocarbon. The doublet in the 1380 cm^{-1} range shows a much stronger lower line, indicating the probability of a *tert*-butyl group. This is confirmed by the presence of multiple bands in the fingerprint region. There is additional fine structure to the *gem*-dimethyl doublet, indicating that there is another dimethyl group present, and there are no lines indicating propyl or ethyl groups in the 750–800 cm^{-1} range. These data are consistent with only one compound: iso-octane $(CH_3)_3CCH_2CH(CH_3)_2$.

Problem 2.3 (C_7H_5N)

The strong, sharp peak at 2225 cm^{-1} is a clear indication of a nitrile group. Since subtracting CN from the formula leaves C_6H_5, a phenyl group might be inferred. This is borne out by the strong peaks at 690 and 760 cm^{-1}, which along with the four peaks between 1650 and 2000 cm^{-1} imply a mono-substituted benzene. The compound is benzonitrile.

Problem 2.4 ($C_4H_{10}O$)

The large band at 3350 cm^{-1} is clearly indicative of an alcohol. The doublet at 1380 cm^{-1} shows that a *gem*-dimethyl group is present and that it is *not* *tert*-butyl. Only iso-butyl alcohol and *tert*-butyl alcohol might be considered possibilities. However, the band at 1065 cm^{-1} is too low for 2° or 3° alcohols and the compound is thus $(CH_3)_2CHCH_2OH$.

Problem 2.5 $(C_6H_{15}N)$

The compound is clearly saturated and might well be an amine. Since there are no strong, broad peaks above 3200 cm^{-1}, we can deduce that it must be tertiary. The peak at 790 cm^{-1} is indicative of an ethyl group, and since there are no peaks in the proper range for a propyl group (734–743 cm^{-1}), we can conclude that the compound is triethylamine.

Problem 2.6 $(C_8H_8O_2)$

The compound is clearly aromatic from the line at 710 cm^{-1} and the formula. Furthermore, it contains a carbonyl from the peak at 1725 cm^{-1}. It is not a carboxylic acid from the absence of OH absorptions, and in such a small compound the number of carbonyl-containing species is limited to methoxy-substituted benzaldehydes, phenylacetaldehydes, and methyl benzoate. The carbonyl must, in all cases, be in conjugation with the ring and could only come at such a high frequency in methyl benzoate, where the methoxy group increases the stretching frequency of the carbonyl. This is further confirmed by the peak at 710 cm^{-1}, characteristic of mono-substituted benzenes, and the absence of an aldehyde peak at 2720 cm^{-1}.

Problem 2.7 $(C_9H_{10}O)$

The strong absorption at 1685 cm^{-1} indicates a carbonyl group in conjugation with the aromatic ring. The lines at 3040 and 820 cm^{-1} confirm the presence of such a ring. The pattern between 1650 and 2000 cm^{-1} and the band at 820 cm^{-1} indicate a *para*-substituted ring. The compound is thus *para*-methylacetophenone.

Problem 2.8 (C_6H_6NCl)

The peaks at 750 and 690 cm^{-1}, as well as those at 3070 and 3020 cm^{-1}, indicate an aromatic compound. The 3380–3470 cm^{-1} doublet is indicative of a 1° amine and the pattern at 1650–2000 cm^{-1} is characteristic of *ortho* substitution. The compound is *o*-chloroaniline.

Problem 2.9 $(C_7H_8O_2)$

The band at 3420 cm^{-1} is indicative of a hydrogen-bonded alcohol and those at 1050 and 1035 cm^{-1} of the C—O stretch. The band at 1185 cm^{-1} and that at 2850 cm^{-1} indicate the presence of a methoxy group, leading to the likelihood of a methoxyphenol. The band at 755 cm^{-1} and the pattern between 1650 and 2000 cm^{-1} indicate *ortho* substitution. The compound is *ortho*-methoxyphenol or guiaicol.

Problem 2.10 (C$_7$H$_5$OCl)

The peaks at 1700 and 2775 cm^{-1} are indicative of an aromatic aldehyde. The aromaticity is confirmed by the formula and the strong peaks below 800 cm^{-1}. The pattern between 1650 and 2000 cm^{-1} is sufficiently visible to predict *ortho* substitution. The compound is *o*-chlorobenzaldehyde.

Problem 2.11 (C$_6$H$_{10}$O)

This is a most unusual spectrum. The formula indicates a hydrogen deficiency of 4, and while there are C—H stretches above 3000 cm^{-1}, it is apparent that there are none of the expected C=C stretches in the 1600 cm^{-1} range. The compound is clearly not aromatic. It therefore must contain two rings and no double bonds. The band at 3090 cm^{-1} could be indicative of either an epoxide or a cyclopropane. However, the bands at 1255 and 835 cm^{-1} are common in epoxides and the compound is thus cyclohexene oxide.

Problem 2.12

(a) There could be two different kinds of carbonyl groups. (b) There could be β-diketone conjugation. (c) It could be a dilute carboxylic acid showing both the monomer and dimer peaks. (d) There could be Fermi resonance. (e) There could be vibrational coupling between two carbonyl groups. (f) There could be partial chelation with a nearby OH group.

Problem 2.13

The two bands in ethanol, propanol, and isobutanol can be considered to have normal intensity ratios for primary alcohols. However, the substantial difference in the ring case of cyclopropylmethanol may be attributed to —OH interaction with the cyclopropyl ring, rather like a π-bond. Similar effects have been noted in allyl and benzyl alcohol.

Problem 2.14

The carbonyl frequency in ethyl benzoate would be in the 1720 cm^{-1} range, lowered by the aromatic ring, but raised by the ethoxy group. The carbonyl frequency in phenyl acetate would be at least 10 cm^{-1} higher since it is not in conjugation with the ring. Furthermore, we would expect to see an ethyl rocking frequency around 790 cm^{-1} in ethyl benzoate, but not in phenyl acetate.

Problem 4.1 ($C_{14}H_{14}$)

If the compound were saturated and contained no rings, it would have the formula $C_{14}H_{30}$. There are therefore $(30 - 14)/2 = 8$ rings + double bonds. Since there are no vinyl protons and there is an aromatic singlet, we assume that there are two benzene rings ($C_{12}H_{10}$). This leaves only C_2H_4 corresponding to the upfield peak at $\delta 2.89$. A peak at this position is most probably due to benzylic hydrogens, and since all four of them are equivalent, they must be symmetrically arranged. Two possibilities might be 4,4′-p-dimethylbiphenyl, CH_3—Ph—Ph—CH_3, and bibenzyl, $PhCH_2CH_2Ph$. Looking back to the spectra of toluene and p-xylene in Figure 4.1 we discover that methyl groups attached to benzene rings come higher, at $\delta 2.35$. Thus we probably have two methylene groups attached between the rings, and the compound is bibenzyl. This is confirmed by the apparent integration, where there are clearly more aromatic than aliphatic protons.

Problem 4.2 (C_8H_9Br)

This is also clearly an aromatic compound having a single benzene ring, which can be seen from the formula and the multiplet at $\delta 7.16$. The upfield triplet and quartet indicate an ethyl group, and the deshielding of the methylene protons to $\delta 2.55$ indicates that they are probably attached to the benzene ring. This leaves only the problem of positioning the bromine. It cannot be on the ethyl group because of the multiplicity, and from the characteristic pair of doublets in the aromatic region we can unambiguously say that it is *para* to the ethyl group. The compound is p-bromoethylbenzene.

Problem 4.3 ($C_4H_6Cl_4$)

Despite the triplet at $\delta 4.6$, the formula clearly indicates that the compound is saturated. We must thus arrange the four carbons and four chlorines to account for the simple pattern we observe. There are only two kinds of protons, and since the areas of the two multiplets are not equal, they are presumably in a ratio of 4:2. They must further be arranged so that two of one kind are adjacent to one of the other kind to produce the observed doublet and triplet. There is only one possible structure with these features: 1, 2, 3, 4-tetrachlorobutane. The protons on carbons 2 and 3 are equivalent and adjacent to each other, so they are split only by the protons on carbons 1 and 4.

Problem 4.4   ($C_7H_{14}O$)

There is one double bond or ring in this compound from the formula. However, the compound is clearly not vinyl, as can be seen from the shifts. It must be highly symmetrical to give so few peaks, which would eliminate most ring compounds. It is not an alcohol or an ether from the chemical shifts, and thus must contain a carbonyl group to account for the double bond. While the peaks at $\delta2.2$ may appear to be a doublet at first, they cannot be since there is no other splitting to indicate that they are coupled to something. The spectrum thus consists of three singlets whose *intensities* are apparently in the ratio of about 3.5:6:18. (The integrals must be similar since the line widths are the same.) The large peak at $\delta1.03$ is several equivalent methyl groups, and if we assume there are three such groups, we have five protons left to assign. These are probably three of one kind and two of another. We thus have three methyls of one kind, and another three and two protons as well as a carbonyl to arrange. Three equivalent methyls in a compound of this size clearly means a *tert*-butyl group, which takes four of the seven carbons. The remaining three are the CH_2, the $C=O$, and the CH_3. Since none of these is coupled, the carbonyl must separate them. The compound is $(CH_3)_3C-CH_2-CO-CH_3$, 4,4-dimethyl-2-pentanone.

Problem 4.5   ($C_4H_6Cl_2$)

The formula suggests that there is one ring or double bond. Since it contains only four carbons and has peaks at $\delta5.6$, we assume that there is one double bond. There are no complex vinyl splittings so the compound must be fairly symmetrical. The singlet at $\delta2.18$ suggests a methyl group which is uncoupled to any other proton but is adjacent to a double bond. The methyl might be uncoupled if the other substitutent on the double bond is one of the chlorines. This suggests the fragment $(CH_3)ClC=$. The splitting of a doublet and triplet in the vinyl region indicates one proton coupled to two others. This suggests the fragment $=CH-CH_2Cl$. The compound is thus $(CH_3)ClC=CHCH_2Cl$, 1,3-dichloro-2-butene.

Problem 4.6   (C_4H_8O)

This is a deceptive spectrum at first. The formula indicates one double bond or ring, and a double bond is selected because of the doublet at $\delta4.85$. However, this "doublet" gives the impression of some spin coupling to another group when there is no other splitting observed. The broadened singlet at $\delta2.2$ is presumably an OH proton and the compound is thus probably an alcohol. The tall peak at $\delta1.75$ must be a methyl group, deshielded by being attached to a double bond, the singlet at $\delta 4.0$ probably a CH_2 attached to the OH. We must arrange these fragments so as to give the observed spectrum.

The singlet at $\delta 4.0$ could not be deshielded as much as it is because of the oxygen and it must also be attached to the double bond. Thus the possible structures are $(CH_3)CH=CHCH_2OH$ and $CH_2=C(CH_3)CH_2OH$. Regardless of whether the groups are arranged *cis* or *trans*, we would expect substantial amounts of coupling between the various protons in the first of these, the 2-buten-l-ol, but might observe only small, long-range coupling for the second of these, the 2-methyl-2-propen-l-ol (methallyl alcohol). The doublet at $\delta 4.85$ can thus be rationalized as two different protons on the same carbon coupled to each other. However, since the geminal vinyl coupling constant is usually only 0–2 Hz and these peaks are separated by 7 Hz, we conclude that we have simply two overlapping singlets coupled only slightly, leading to their slightly broadened character. Little long-range coupling is observed in this molecule, perhaps because the electronegativity of the oxygen makes the electrons less available for transmission of coupling information. The compound is methallyl alcohol.

Problem 4.7   (C_8H_8)

The formula indicates five rings or double bonds. There is obviously one aromatic ring present and one additional double bond. The characteristic *ABX* pattern in the region of $\delta 5.0–\delta 7.0$ indicates an $R—CH=CH_2$ compound, and this can only be styrene, $Ph—CH=CH_2$.

Problem 4.8   (C_5H_9ClO)

This compound has one double bond from the formula and appears to be entirely first order, consisting of four kinds of CH_n groupings, three of them adjacent to CH_2's and one adjacent to a CH_3. The larger triplet at $\delta 1.08$ is the methyl and it must be adjacent to a CH_2 which is split into a quartet, at $\delta 2.45$. There is no evidence of an $—OH$ singlet for an alcohol, nor are there any vinyl protons that would have to occur to explain the double bond if the compound were an ether. It is therefore a ketone with the carbonyl separating the methylenes to maintain the simple spectrum. The triplet at $\delta 3.72$ is the methylene attached to the chlorine and the compound is l-chloro-3-pentanone.

Problem 4.9   ($C_3H_5ClO_2$)

This compound has one double bond and it is clear from the far downfield peak that it is a carboxylic acid. The remaining protons and chlorine must be arranged on the other two carbons so that a doublet and a quartet will be produced: one of one kind and three of the other. This can only be obtained by an α-chlorocarboxylic acid. The compound is 2-chloropropionic acid.

Problem 4.10   ($C_{10}H_{14}$)

This is a mono-substituted aromatic compound from the formula and the downfield aromatic singlet. This predicts the fragment C_6H_5 which leaves behind C_4H_9, a saturated butyl group of some kind. The only possible compounds are *n*-butylbenzene, iso-butylbenzene, *sec*-butylbenzene, and *tert*-butylbenzene. The *tert*-butyl compound is eliminated because it would have only an upfield singlet. The iso-butyl compound is eliminated because we would expect to see the characteristic isopropyl pattern of a strong doublet and a weak septet, which we do not observe, although a doublet and some multiplets are present having the wrong intensity ratios. We can eliminate the *n*-butyl case because we would expect to see a downfield benzylic triplet, and the peaks in the benzylic position appear here to be a sextet. This leaves only the *sec*-butylbenzene as a possible structure. The sextet at $\delta 2.55$ is caused by the overlapping splittings from the methyl and methylene groups adjacent to the Ph—CH group, the doublet by a CH_3 next to the methine CH, and the triplet and quartet that remain by the ethyl group. The compound is thus CH_3—CH_2—$CH(CH_3)Ph$, *sec*-butylbenzene.

Problem 4.11   (C_8H_8O)

From the formula, we see that there must be five double bonds and/or rings. The downfield singlet indicates a mono-substituted aromatic ring having the formula C_6H_5. This leaves C_2H_3O to arrange. Neither an aldehyde (Ph—CH_2—CHO) nor a ketone (Ph—CO—CH_3) would give the requisite shifts and splittings and a vinyl alcohol is not possible. The compound thus must be an epoxide. This is further justified by the characteristic *ABX* pattern between $\delta 2.5$ and $\delta 4.0$. It is further split in this case by long-range couplings from the aromatic ring. The compound is styrene oxide.

Problem 4.12   ($C_{11}H_{16}O$)

From the formula and the downfield pair of doublets this must be a *para*-substituted benzene ring. The presence of oxygen in the formula and the broad singlet at $\delta 4.62$ indicates an alcohol. The sharp singlet at $\delta 1.25$ indicates at least two or three equivalent methyl groups that are not benzylic, and the triplet at $\delta 0.70$ is undoubtedly another methyl that must be coupled to the broadened multiplet at $\delta 1.62$, which therefore must be a quartet. We thus must arrange the fragments X—Ph—Y, CH_3CH_2, —OH, and CH_3—C—CH_3 to account for this spectrum and formula. We have exactly 11 carbons listed among these fragments, which eliminates a *tert*-butyl group instead of the CH_3—CH—CH_3 group. Furthermore, there are probably not carbons on each end of the benzene ring or the *para* doublets would not be so separated. The only other possible group is an —OH and the compound is thus *p-tert*-amylphenol, $CH_3CH_2C(CH_3)_2$—Ph—OH.

Problem 4.13   ($C_5H_8F_4O$)

This compound is actually easier to analyze than might first appear since fluorine splitting patterns with protons are nearly always first order. The triplet of triplets between 5 and 7 ppm is thus a proton split widely by two very near fluorines and split more narrowly by two more distant ones. This suggests the group HCF_2CF_2—immediately. The large, sharp peak at $\delta 1.41$ is at least two methyl groups, and they must be close enough to the fluorines so that they are slightly split by long-range couplings. This suggests the structure HCF_2—CF_2—$C(CH_3)_2$. Counting up, we find that this leaves only the elements OH which account for the singlet at $\delta 2.1$. The compound is thus 2-methyl-3,3,4,4-tetrafluoro-2-butanol.

Problem 4.14   ($C_{10}H_{10}ClNO_2$)

While this may look like an imposing formula, the spectrum is extremely simple and can be easily analyzed. There is clearly a *para*-substituted benzene ring with substantially different groups involved. This suggests the fragment Cl—Ph—R, where we now have elements $C_4H_6NO_2$ left for R. Since there are only three singlets left to assign, this is not too difficult to guess. There are two double bonds implied in the formula $C_4H_6NO_2$, and since there are no peaks in the vinyl region, these must be carbonyls. If we subtract our two C=O's, we have left C_2H_6N, which could produce three singlet peaks only if they were CH_2, CH_3, and NH separated by the carbonyls. The possibility of fragments CH_2, CH_3, and NH_2 is eliminated by the relative intensities. We can arrange these fragments, then, as CH_2—CO—NH—CO—CH_3 or —NH—CO—CH_2—CO—CH_3. We cannot choose between these easily, but if we examine the chemical shift of the peak at $\delta 3.55$, we might decide that since it is the CH_2, the shift would be further downfield if it was between the benzene ring and the carbonyl than if it was between the two carbonyls. The compound is *p*-chloracetoacetanilide, p—Cl—Ph—$NHCOCH_2COCH_3$.

Problem 5.1

If the computer is to observe a spectral width of 4200 Hz, then it must sample at a rate of 8400 Hz. The sampling rate will then be 1/8400 or 119 μsec point.

Problem 5.2

^{31}P will resonate at (36.44 MHz) 200/90 = 80.978 MHz. At this frequency, 25 ppm will be 2024 Hz and the sampling rate will have to be 1/2 (2024) = 247 μsec/point.

Problem 5.3

The answer to this is yes and no. When quadrature detection is used, the rf carrier is set in the middle of the spectrum instead of at one end, and data are obtained from a spectral width half as wide on either side of the carrier. At first, it might seem that this would make it possible to observe spectra over twice the spectral width as under single-detector FT schemes, but since two points must be taken, one from each detector, the net result is that the overall sampling rate will be the same in terms of the analog-to-digital converter duty cycle.

Problem 5.4

It is true that quadrature detection will avoid foldback of peaks below the carrier frequency, but only as long as they are within the bandwidth defined by the sampling rate. For example, to observe a 5000 Hz spectral width, we conventionally place the rf carrier in the center of the region of interest and observe 2500 Hz on either side of the carrier. If peaks occur outside this 2500 Hz range on either side of the middle they *will* fold back.

Problem 6.1 (C_8H_{18})

From the spectrum and the formula this is obviously a saturated hydro-carbon. It contains two different kinds of methyl groups from the splitting information (25.5q and 30.2q), one CH (24.9d), one CH$_2$ (53.4t), and one quaternary carbon (31.2s). Since there are eight carbons to account for and since the line at 30.2 is most intense, we might predict that we have the fragment $(CH_3)_3C$—, a *t*-butyl group. There are, then, only three other kinds of carbons, CH$_2$, CH, and CH$_3$, and there must be four carbons among them, so we arrive at the fragment CH$_2$—CH—(CH$_3$)$_2$ to explain this. These can only be combined to form iso-octane, $(CH_3)_3C—CH_2—CH(CH_3)_2$. The calculated assignments from the Lindeman-Adams equation are:

	Calculated	Experimental
$C_1 = 6.80 + 25.48 − 2.99 + 0.49$	29.79q	30.2q
$C_2 = 27.77 + 2.66 + 2(0.86) =$	30.81s	31.2s
$C_3 = 15.34 + 21.43 + 16.70 =$	53.47t	53.4t
$C_4 = 23.46 + 6.60 + 3(−2.07) =$	23.85d	24.9d
$C_5 = 6.80 + 17.83 − 2.99 + 3(0.49) = 23.11q$		25.5q

The assignments thus are:

$$(CH_3)_3—C—CH_2—CH—(CH_3)_2$$
$$30.2 \quad 31.2 \quad 53.4 \quad 24.9 \quad 25.5$$

Problem 6.2 $(C_5H_{11}Cl)$

This is obviously a chlorinated hydrocarbon containing a CH_3, a CH, and two CH_2's. Since there is only one kind of methyl group present, and the CH_3 peak is so intense, it must represent two CH_3's. This also explains the CH and allows two CH_2's. The only possible structure is $(CH_3)_2—CH—CH_2—CH_2—Cl$, 1-chloro-3-methylbutane. The calculated line assignments are obtained by calculating the line positions for 2-methyl-butane by the Lindeman-Adams rules and then adding in the parameters for Cl substitution from Table 6.7. These are:

	RH		add R—Cl	Experimental
$C_1 = 6.80 + 17.83 − 2.99 =$	21.64		21.64q	22.0q
$C_2 = 23.46 + 6.60 =$	30.06	− 4	26.06d	25.7d
$C_3 = 15.34 + 16.70 =$	32.04	+ 11	43.04t	43.1t
$C_4 = 6.80 + 9.56 + 2(−2.99) = 10.33$	+ 31		41.33t	41.6t

Problem 6.3 $(C_4H_8O_2)$

This compound clearly contains a carbonyl, as can be seen from the peak at 170.7 and from the double bond or ring present, given by the formula. The other oxygen must be saturated, and this suggests a carboxylate group COO. There are two methyls and one methylene from the splittings and the compound must therefore be ethyl acetate: $CH_3COOCH_2CH_3$. The lines can be assigned by subtracting the values in Table 6.6 from the chemical shifts of the two alkyl fragments. We consider the molecule as CH_3R and CH_3CH_2R', where $R = COOEt$ and $R' = CH_3COO$. The calculation proceeds as follows:

Parent alkane	Shift	R-Group	Amount to add	Calculated	Experimental
CH_4	-2.1	COOEt	20	18.1q	20.9q
CH_3—CH_3	5.9	CH_3COO—	51 (α)	56.9t	50.4t
CH_3—CH_3	5.9	CH_3COO—	6 (β)	11.9q	14.4q
$C{=}O$					170.7

The lines are thus assigned:

$$CH_3—COO—CH_2—CH_3$$
$$20.9 \quad 170.7 \quad 60.4 \quad 14.4$$

Problem 6.4 (C_4H_7Br)

From the chemical shifts and the formula, the compound is clearly an alkene. The fragments presently include CH_3, two CH's, and a CH_2 from the multiplicities. These can be arranged in two ways to include the Br as follows: CH_3CH_2—$CH{=}CHBr$ and CH_3—$CH{=}CH$—CH_2Br. The calculated shifts for the vinyl carbons in the two structures are:

1-bromo-1-butene	1-bromo-2-butene
$C_2 = 123.3 + 10.6 + 7.2 - 1 = 140.1$	$C_2 = 123.3 + 10.6 + 0 - 7.9 = 126$
$C_1 = 123.3 - 8 - 7.9 - 1.8 = 105.6$	$C_3 = 123.3 + 10.6 - 7.9 + 2 = 128$

From these data alone, the compound cannot have a bromine on a vinyl carbon. The compound is 1-bromo-2-butene, and the assignments are $C_1 = 32.9$, $C_2 = 127.8$, $C_3 = 131.0$, and $C_4 = 17.5$.

Problem 6.5 (C_6H_8O)

This compound has three double bonds or rings in it, and both a carbonyl and a methyl group are present. Since the carbonyl carbon is split into a doublet (193.0d) this must be an aldehyde. Since there are no singlets nor the correct chemical shifts there can be no triple bonds, and the only question is whether the double bonds are in conjugation with the carbonyl and with each other. If they are not in conjugation with both the carbonyl and each other, there cannot be a terminal methyl group (18.7q) and thus the structure must be 2,4-hexadienal. The assignments of carbons are:

Calculated	Experimental
$C_2 = 123.3 - 7.9 - 1.8 + 1.5 + 13 = 128.1$	130.4 or 130.2
$C_3 = 123.3 + 10.6 + 7.2 - 1.5 + 13 = 152.6$	152.3
$C_4 = 123.3 + 10.6 + 7.2 - 7.9 = 133.3$	141.6
$C_5 = 123.3 + 10.6 - 7.9 - 1.8 = 124.2$	130.4 or 130.2
$C_1 =$	193.0
$C_6 =$	18.7

Problem 6.6 (C_9H_{12})

The presence of four rings or double bonds in the formula seems to indicate an aromatic ring. This is further justified by the compound's extreme symmetry, having only three kinds of carbons. Since one kind is a methyl group (21.2q) the compound must be 1,3,5-trimethylbenzene, mesitylene. Without any calculations, we can assign the unsplit, low NOE line (137.5s) to the methyl-substituted carbons (1, 3, 5) and the δ127.2d to carbons 2, 4, and 6.

Problem 6.7 ($C_6H_{15}NO$)

Since the compound has six CH_2 groups from the six triplets, it must be a 1,5-disubstituted hexane. The remaining elements make it 6-amino-1-hexanol. Assignments can be made by calculating carbons 1, 2, and 3 for hexane and then adding on the parameters from Table 6.6 for OH and HN_2.

	Alkane	Additional values	Calculated	Experimental
$C_1 = 6.80 + 9.56 - 2.99 + 0.49 =$	13.86	+29(OH)	42.86	43.2
$C_2 = 15.34 + 9.75 - 2.69 + 0.25 =$	22.65	+11	33.65	34.9
$C_3 = 15.34 + 2(9.75) - 2.69 =$	32.15	−5	27.15⎫	27.0 and 27.9
$C_4 =$	32.15	−5(NH_2)	27.15⎭	
$C_5 =$	22.65	+10	32.65	34.2
$C_6 =$	13.86	+48	61.86	62.9

The assignments of carbon 1, 2, 5, and 6 are definite and those of carbons 3 and 4 can be interchanged.

Problem 6.8 ($C_9H_{10}O$)

This compound contains a carbonyl and is most probably aromatic as well. Since it also contains an ethyl group (8.2q, 31.6t), we have identified all the nine carbons and need only arrange them. Since the carbonyl is unsplit (200.0s) it is not an aldehyde, and since there are four kinds of aromatic carbons, this must be a mono-substituted benzene. The compound is CH_3CH_2COPh, propiophenone. By inspection we can assign the carbonyl and ethyl group, and the peak with least NOE must be C_1 of the ring (137.2s). The C_4 ring carbon similarly must be less intense since there is only one of them (132.8d). The $C_{2,6}$ and $C_{3,5}$ peaks are too close to assign and could be either 128.0 or 128.6.

Problem 6.9 ($C_5H_8O_2$)

The compound contains a carbonyl (166.0s), a vinyl group (129.3t, 130.0d), and an ethyl group (60.4t, 14.4g). There is clearly a terminal methylene (CH_2CH) from the vinyl triplet (129.3t). The compound could be ethyl propenate ($CH_2CHCOOCH_2CH_3$) or allyl propanoate (CH_2=$CHOOCCH_2CH_3$). These can be distinguished by the downfield shift of the alkyl CH_2 (60.4t), and the compound is in fact ethyl propenoate. This is confirmed by shift calculations:

	Alkyl	Additional values	Calculated	Experimental
CH_3	5.9	+6	15.9q	14.4
CH_2	5.9	+51	56.9t	60.4
=CH	123.3	+6	129.3d	130.0
CH_2=	123.3	+7	130.3t	129.3

Note the interchange of the calculated vinyl shifts with the experimental ones to account for the off-resonance multiplicities.

Problem 6.10 ($C_9H_{10}O_3$)

This is an aldehyde (190.7d) from the carbonyl doublet and some sort of substituted benzene from the six carbons in the 115–151 range. The two low-field methyl groups (56.0q, 56.1q) can only be explained by two methoxy groups. Since all six aromatic carbons are different, there must be no symmetry, which eliminates all but a few of the dimethoxy benzaldehydes. By calculation using Table 6.4, we can arrive at shifts for 2,3-dimethoxy-benzaldehyde and 3,4-dimethoxybenzaldehyde, the only asymmetric possibilities:

C	2,3	2,4	3,4	Experimental
1	123.7s	123.7s	130.4s	130.3s
2	146.8s	153.5s	116.4d	110.7d
3	146.1s	115.7d	146.1s	149.8s
4	120.6q	120.6s	151.0s	154.6s
5	122.4d	152.8d	115.7d	109.4d
6	123.1d	116.4d	123.1d	126.5d

Careful inspection allows us to conclude that the only compound showing two upfield and two downfield peaks of the correct multiplicity is the 3,4-dimethoxybenzaldehyde, whose assignments are given correctly in the last column.

Problem 6.11 ($C_{12}H_{14}O_4$)

This must be a highly symmetric compound since it shows only six lines for 12 carbons. It clearly contains a carbonyl (167.5s), an ethyl group (14.2q, 61.5t) and is probably aromatic. Since the intensities of all the nonquaternary carbons are roughly equal, all the carbons must be doubled in the compound. Since the methylene group is so far downfield (61.5t), it is probably attached to an oxygen. Furthermore, since the carbonyl is so far upfield, it also must be attached to an oxygen. This suggests the $COOCH_2CH_3$ functional group. A totally symmetrical compound having these features is diethyl phthalate, $o\text{-Ph(COOEt)}_2$. The calculated and assigned lines are:

	Calculated	Experimental
CH_3		14.2q
CH_2		61.5t
$C{=}O$		167.5s
$C_{1,2} = 128.5 + 2.1 + 1.1 = 131.7s$		132.7s
$C_{3,6} = 128.5 + 1.1 + 0.1 = 129.7d$		129.0d
$C_{4,5} = 128.5 + 0.1 + 4.5 = 133.1d$		131.1d

Problem 6.12 ($C_9H_8O_3$)

The carbonyl is far enough upfield that it must have an oxygen attached, and since there are no attached alkyl groups, we may assume the COOH functional group. From the formula, this compound has six rings and double bonds. There are five remaining after assigning the carbonyl. The symmetry of the compound suggests aromaticity and this eliminates another four rings and double bonds. There must be another double bond on the aromatic ring as well. We now have explained all but the third oxygen, which must be another OH group somewhere on the structure. If two of the vinyl region carbons are part of a double bond, then there are only four kinds of aromatic carbons, suggesting *para* substitution. If the COOH is on one end of the ring, this leaves the components of a vinyl alcohol at the other end. This is not a possible structure, and we then consider putting an OH on one end of the ring and the vinyl group on the other end, with the COOH attached to it. This leads to the structure 4—HO—Ph—CH=CH—COOH, 4-hydroxycinnamic acid, which is the correct structure. The calculated and assigned shifts are:

	Calculated		Experimental
C_1	$= 128.5 + 9.5 - 7.3 =$	130.7s	125.4s
$C_{2,6}$	$= 128.5 - 2.0 + 1.4 =$	127.9d	130.0d
$C_{3,5}$	$= 128.5 - 12.7 + 0.2 =$	116.0d	115.4d or 115.9d
C_4	$= 128.5 + 26.9 - 0.5 =$	154.9s	159.7s
C_α	$= 123.3 + 12 + 9$	144d	144.2d
C_β	$= 123.3 - 11 + 4$	116d	115.9d or 115.4d

Problem 6.13 ($C_8H_{11}N$)

The compound contains an ethyl group and six vinyl region hydrogens and is presumably aromatic. This leaves the elements of NH_2. It would be an ethylaniline or an N-ethylaniline. This latter compound is eliminated since all six ring carbons are different. This similarly eliminates p-ethylaniline. Examining the calculated shifts for o-ethylaniline and m-ethylaniline we find that the better fit by far is with the o-ethylaniline.

m-Ethylaniline	o-Ethylaniline	Experimental
C1 = 146.5s	C1 = 146.1s	144.3s
C2 = 114.8d	C2 = 130.8s	128.0s
C3 = 144.1s	C3 = 128.1d	128.4d
C4 = 118.3d	C4 = 118.7	118.6d
C5 = 128.5d	C5 = 125.9d	126.8d
C6 = 112.6d	C6 = 115.2d	115.4d

Problem 6.14 (C₇H₈)

This compound must be quite symmetric from its carbon spectrum, since there are only three kinds of carbons. Furthermore, they are olefinic (or aromatic) as well as alphatic. From the IR spectrum, it could be aromatic from the peaks below 800 cm^{-1}, but since it shows no substitution pattern at all between 1650 and 2000 cm^{-1}, it probably is not. The pmr spectrum shows three closely spaced triplets which indicates that all of the protons are adjacent to each other. They could be in pairs or, more likely, they could be two different splittings causing a doublet of doublets to become a triplet. From the carbon spectrum we have an olefinic CH, an alkyl CH, and an alkyl CH$_2$. Since the olefinic CH is so much more intense than the alkyl CH even though their NOE's might be similar, we can assume that there are more olefinic CH's. Olefinic CH's must come in pairs, so that we might postulate 4 =CH's, 2 CH's, and one CH$_2$, the latter being more intense because of its increased NOE's. Thus we could have only a symmetrical olefinic structure such as 1,4-cyclohexadiene with an additional bridge to account for the seventh carbon. The compound is norbornadiene or bicyclo-[2.2.1]-heptadiene.

Problem 6.15 (C₅H₁₁N)

The formula indicates that an amine is present. The singlet at 3300–3500 cm^{-1} indicates a secondary amine. The pmr triplet at $\delta 2.80$ indicates that the group next to the one attached to the nitrogen is a CH$_2$, and the carbon spectrum shows that there must be only three kinds of carbons, all CH$_2$ groups. Since this must be a secondary amine, this indicates a CH$_2$NHCH$_2$ linkage which must be symmetrical. None of the three spectra indicate any olefinic carbons, and thus we conclude that there must be a symmetrical ring structure. The compound is piperidine.

Problem 7.1

(a) Same as butadiene.

(b)

$E = 10\alpha + 13.68\beta.$

(c)

$E = 8\alpha + 10.94\beta.$

(d)

$E = 4\alpha + 5.12\beta.$

(e)

$E = 10\alpha + 12.68\beta.$

Bond orders	
1, 2	1.72
2, 3	1.60
3, 4	1.72
4, 5	1.55
5, 6	1.52
1, 6	1.55
6, 7	1.55
7, 8	1.72
8, 9	1.60
5, 10	1.55
9, 10	1.72
1, 2	1.72
1, 3	1.58
2, 3	1.58
3, 4	1.44
4, 5	1.57
5, 6	1.69
6, 7	1.61
4, 8	1.57
7, 8	1.69
1, 2	1.49
1, 3	1.62
2, 3	1.49
1, 4	1.49
3, 4	1.49
1, 2	1.43
2, 3	1.64
2, 9	1.37
1, 10	1.66

Problem 7.2

Apparently, the additional two electrons cause additional stability in the α-levels so that they become bonding levels instead of nonbonding ones.

Problem 7.3

Cation $2\alpha + 2.83\beta$.
Anion $4\alpha + 2.83\beta$.

Problem 7.4

The reciprocals of the wavelengths, the wave *numbers*, are linear in energy and can be correlated with the calculated bond orders.

Problem 8.1

	Calculated	Observed
(a)	232	230
(b)	227	227
(c)	237	236
(d)	247	245

Problem 8.2

	Calculated	Observed
(a)	244	244(log ε = 4.0)
(b)	273	275(log ε = 3.8)
(c)	244	234(log ε = 4.1)
(d)	280	283(log ε = 4.3)

Problem 8.3

Structure II is correct. The calculated values are II (234) and III (273).

Problem 8.4

	Observed	Calculated
IV	247	$215 + 10 + 12 + 12 = 249$
V	241	$215 + 10 + 12 + 5 = 242$

Problem 8.5   (C_6H_{10})

From the molecular formula this compound has two sites of unsaturation and from the UV spectrum, it must obviously be a diene with two alkyl groups substituted on it [217 + 2(5) = 227, observed 226]. While there seems to be only one kind of methyl group at 1.92, there is a clear doublet at 5.0. This doublet cannot be due to coupling since there is no other splitting, and thus must be due to two kinds of vinyl hydrogens. Since there is no coupling, the methyl groups must be distant from the vinyl hydrogens. The only such compound possible is 2,3-dimethyl-1,3-butadiene. The two kinds of vinyl hydrogens occur because two are *cis* and two are *trans* to the methyl groups.

Problem 8.6   (C_7H_{12})

This compound is also clearly a diene having two sites of unsaturation. From the UV maximum, it has three alkyl substituents (217 + 3 (5) = 232), and they must be arranged so that there is no coupling with adjacent methyl groups. The methyl peak appears to be made up of two peaks, one larger than the other, and could well consist of two types of methyl groups in a ratio of 2:1. These factors can only be accounted for if the compound is 2,4-dimethyl-1,3-pentadiene.

Problem 8.7   (C_7H_8)

The pmr spectrum of this compound is a "jungle," showing only that there are some vinyl and some alkyl protons. Integration will not be too helpful in such a closely coupled system, but there appears to be one kind of alkyl proton split by two equivalent protons to form the triplet. The molecular formula indicates four double bonds + rings, and the low UV maximum indicates that there cannot be four double bonds. The most important information is given by the carbon spectrum, however. It indicates that there are only four types of carbons in the compound, three with one proton attached and one with two attached. The only cyclic system that can have these properties is tropylidine, 1,3,5-cycloheptatriene.

Problem 8.8   ($C_6H_{10}O$)

This compound contains two double bonds. The chemical shifts and lack of splittings do not suggest an ether or alcohol. It thus must be an α, β-unsaturated ketone. Since the base for such absorptions is 215 this must have two alkyl groups on the alkene (235 − 215 = 20). The compound contains two types of CH_3 groups, in a ratio of 2:1, and one vinyl proton. This can only be explained by $(CH_3)_2$—C=CH—CO—CH_3, mesityl oxide.

Problem 9.1 ($C_2H_4O_2$)

The mass is 60 and there are clearly no nitrogens. By the ratio of (M + 1)/M there are two carbons. The M + 2 peak suggests two oxygens (0.00204 × 852) × 2 = 3.5 versus 4. Major peaks at M-17 and M-45 suggest a COOH group. The M-15 confirms this. The compound is CH_3COOH, acetic acid.

Problem 9.2 ($C_6H_{12}O$)

The major fragment at 57 has four carbons in it (48/100 = 0.048). The other fragment at 43 is complementary and could be $CH_3C{\equiv}O^+$ or C_3H_7. The small M-15 peak at 85 is more in line with the ketone than the loss of a methyl from a hydrocarbon. The fact that there is little M-29 (81) indicates that the m/e 57 group always leaves all at once, characteristic of a *tert*-butyl group. The peaks at m/e 41 and 29 are "random rearrangements," often observed in *tert*-butyl groups. The compound is methyl-*tert*-butylketone.

Problem 9.3 (C_9H_{12})

The (M + 1)/M ratio suggests 9 or 10 carbons. The m/e is not large enough for 10 and the compound is C_9H_{12}. The peaks at 91 and 65 are characteristic of alkyl benzenes: the tropyllium ion, and the cyclopentadienyl cation. The peak at 92 is suggestive of a type H rearrangement involving a γ-hydrogen. This precludes alkyl branching and the compound is *n*-propyl benzene.

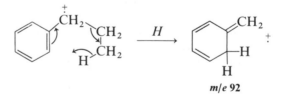

m/e 92

Problem 9.4 ($C_4H_8O_2$)

A parent ion of m/e 88 precludes any possible hydrocarbon. The formula must thus be $C_5H_{12}O$ or $C_4H_8O_2$. The m/e 15 and 24 suggest methyl and ethyl groups. The m/e 43 suggests $CH_3C{=}O$ and m/e 45 OCH_2CH_3. The M-60 peak at m/e 28 suggests elimination of acetic acid, and the mass 61 peak is very common in acetates as a form of protonated acetic acid. The compound is ethyl acetate.

Problem 9.5 (C_7H_{12})

The m/e 96 as a hydrocarbon can only be C_7H_{12}. The large M-15 clearly indicates a branched methyl group, and the large peak at m/e 54 must be due to C_4H_6, butadiene. Since we know this is a cyclic hydrocarbon that has eliminated butadiene, this must be a type D cleavage and the original compound must have been 4-methylcyclohexene.

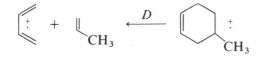

Problem 9.6 (C_2H_5BrO)

From the doublet pattern separated by 2 amu, this must contain bromine. The m/e 45 and 44 peaks confirm this with loss of Br and HBr. The large peak at m/e 31 must correspond to a type B cleavage, where CH_2Br has been lost. This and the M-17 peak at 107–109 confirm the presence of OH. The compound is $HOCH_2CH_2Br$, ethylenebromohydrin.

Problem 9.7 $(C_6H_5NO_2)$

This odd mass parent ion indicates the presence of a nitrogen, and the peaks at 77 indicate an aromatic ring. The peaks at 51, 68, and 93, indicate that this is nitrobenzene with fragments as shown on page 286.

Problem 9.8 $(C_9H_{10}O_2)$

The strong peaks at 65 and 91 indicate an alkyl aromatic system. The large peaks at 118 and 119 correspond to loss of 32 and 31 amu. Since 118 must be an odd electron species, the species lost must be a neutral molecule. This can only be CH_3OH. The remaining groups must be C=O and the compound an aryl ester. The m/e 119 is the loss of OCH_3, but elimination of CH_3OH can only occur by type E_2 rearrangement if the compound is an *ortho* ester. The compound is *ortho*-methyl toluate.

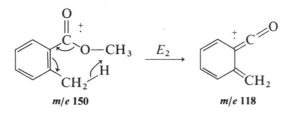

m/e 150 $\qquad\qquad$ m/e 118

Problem 9.9 (C$_4$H$_9$BrO)

The doublets in the mass spectrum indicate the presence of a bromine. Since the molecular weight is only 152, there can only be one such bromine, and the remaining mass to be accounted for is 73. From the decoupled ^{13}C spectrum there are only four kinds of carbons (the 1:1:1 triplet is caused by CDCl$_3$). The proton nmr shows no evidence of unsaturation nor does the IR, so there must be nine hydrogens associated with these four carbons (subtracting 1 from 10 for the Br). Subtracting the fragment C$_4$H$_9$ ($m = 57$) from the remaining mass (73) we get an unaccounted for mass of 16. We suspect that this must be an oxygen since there are no other carbons and there is a strong stretch around 1200 cm^{-1} in the IR. The strong peak at m/e 59 in the mass spectrum is a typical ether peak caused by loss of a CH$_2$Br group to form CH$_3$CH$_2$O=CH$_2$. This is further confirmed by type G cleavage to m/e 31, CH$_2$=OH$^+$. Furthermore, there are no OH or C=O lines in the IR for other possible oxygen groups. The proton nmr does not help us very much, although we can deduce from the upfield triplet that there is a CH$_3$ with an adjacent CH$_2$ group.

From the ^{13}C off-resonance decoupled spectrum we find that there are four types of carbons, three bearing two hydrogens and one bearing three hydrogens. The compound is thus CH$_3$CH$_2$OCH$_2$CH$_2$Br, 2-bromoethyl ethyl ether.

Problem 9.10 (C$_8$H$_9$NO)

From the proton nmr spectrum, this is clearly a *para*-substituted benzene. The odd mass number suggests the presence of one nitrogen, and the doublet at 330 cm^{-1} suggests a primary —NH$_2$. This is further confirmed by the broad resonance at $\delta 4.4$. The peak at 1650 cm^{-1} suggests the presence of a carbonyl in conjugation, which is further confirmed by the ^{13}C peak below $\delta 190$. The compound is p-aminoacetophenone, NH$_2$—C$_6$H$_4$—CO—CH$_3$.

Problem 9.11 (C$_9$H$_8$O)

From the IR and cmr, a carbonyl is clearly present, and from the pmr, this must be an aldehyde. Furthermore, the vinyl peaks in the pmr having the same apparent splitting indicate that there is a double bond in conjugation with the aldehyde. Subtracting the aldehyde from m/e 132 we are left with a mass of 103, corresponding to C$_8$H$_7$. This can only be a double bond in conjugation with the ring since the splitting pattern suggests a monosubstituted benzene. The compound is *trans*-cinnamaldehyde C$_6$H$_5$—CH=CH—CHO.

Problem 9.12   (C$_8$H$_8$O$_3$)

This compound has an acidic —OH proton from the IR and pmr, and has two strongly deshielded carbons at $\delta 161.7$ and $\delta 170.5$. From the multiplicity of the carbon signals, this is a disubstituted benzene, and from the IR and cmr it contains at least one carbonyl group, adjacent to the ring. We have thus identified the fragments C$_6$H$_4$, CH$_3$, and C=O, whose mass totals 119. Subtracting from 152 we have 33 amu unaccounted for as noncarbon res-onances. Subtracting the known —OH group, we then have 16 remaining. This is clearly a third oxygen. The strong deshielding of the CH$_3$ group to $\delta 3.9$ indicates that this must be a OCH$_3$ group. We thus are left with the possible structures

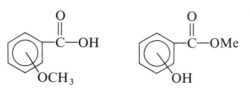

The strong peak at m/e 120 corresponds to loss of CH$_3$OH by a type E_2 elimination,

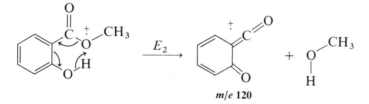

m/e 120

and the compound is thus methyl salicylate

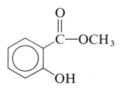

APPENDIX I

LISTING OF
THE LAOCN3 PROGRAM

```
C LAOCOON III
C LEAST-SQUARES ADJUSTMENT OF CALCULATED ON OBSERVED NMR SPECTRA
C BY AKSEL A. BOTHNER-BY AND S.M. CASTELLANO
C
CTHIS PROGRAM CAN CALCULATE THEORETICAL NMR SPECTRA FROM CHEMICAL SHIFTS
C AND COUPLING CONSTANS AND CAN ITERATE THESE THEOREYICAL
C SPECTRA FOR A BEST FIT WITH EXPERIMENTAL LINE FREQUENCIES
C IT IS NOT NECESSARY TO ASSIGN ALL LINES FOR A GOOD FIT TO OCCUR.
C
C MINOR MODIFICATIONS FOR TIME SHARING ON THE DEC-SYSTEM 10
C BY J.W. COOPER 8/78
C ROUTINES FOR PLOTTING ON TEKTRONIX 4012 BY J.W.C.
C ADAPTED FROM THOSE BY R.D. STOLOW AND N. BIRNBERG
C
C
        DIMENSION IL(300),F(300),NAME(15),ISO(7),W(7),A(7,7),IA(29,6),IB(2
     19,6),FZ(128,7),NO(8),LL(8),NP(128),B(300),DC(300,28),X(35,35),D(12
     28,28),VA(35,35),BV(28),VB(35,35),CORR(28),E(128),FNO(5),NKTL(6)
        3 ,IPAR(6),SUM(1024),BUFFER(27)
C THE INPUT ARRAY IPAR AND THE PLOT ARRAY SUM ARE ALSO DIMENSIONED
C
        COMMON NOS,NN,NL,NI,IL,F,NAME,ISO,W,A,IA,IB,FZ,NO,LL,NP,B,DC,X,D,V
     2A,BV,VB,CORR,E,LPT
C DECLARE FILE NAME VARIABLE DOUBLE PRECISION
C SO IT WILL HOLD 10 CHARACTERS
        DOUBLE PRECISION FNAME
C DECLARE TTY SUBROUTINE INTEGER
        INTEGER TTY
C
C THE VARIABLE LPT IS DEFINED AS THE LOGICAL UNIT NUMBER FOR
C THE LINE PRINTER. ON THE DEC-10 THIS IS 3. IT IS COMMONLY 6
C ON MANY OTHER SYSTEMS. BE SURE IT IS ADJUSTED FOR YOUR USE.
C THE VARIABLE HAS ALSO BEEN ADDED TO COMMON TO ALLOW FOR PRINTING
C FROM SUBROUTINES
C
C THIS MAKES THE LPT AN OPTIONAL DISK FILE PRINTED OUT
C ONLY ON REQUEST
        OPEN(UNIT=24,FILE='LPT.TMP',ACCESS='SEQINOUT',DISPOSE='DELETE')
        LPT=24
C
C THE FOLLOWING STATEMENTS CREATE TWO SCRATCH FILES ON DISK NAMED
C TEMP2.TMP AND TEMP3.TMP.  THEY ARE DELETED UPON PROGRAM EXIT
C SINCE THE DEC-10 RECOGNIZES UNITS 20-27 AS DISK FILES, THE UNIT NUMBERS
C HAVE BEEN CHANGED TO 22 AND 23 FROM 2 AND 3. NOTE THAT THIS ALSO REQUIRES
C A CHANGE IN STATEMENTS 24, 25 AND 533.
C
        OPEN(UNIT=22,FILE='TEMP2.TMP',ACCESS='SEQINOUT',
     1 DISPOSE='DELETE',MODE='IMAGE')
        OPEN(UNIT=23,FILE='TEMP3.TMP',ACCESS='SEQINOUT',
     1 DISPOSE='DELETE',MODE='IMAGE')
C THE FOLLOWING CHECKS FOR  THE CORRECT TTY AND INTIALIZES IT TO
C 1200 BAUD-GOOD ONLY AT TUFTS
        TTFLAG=0
        TYPE 9600
```

```
9600      FORMAT(' IS THIS A GRAPHICS DISPLAY TERMINAL? ',$)
          ACCEPT 7111, AQ
          IF(AQ.EQ.'Y')TTFLAG=1
 8005 NNO=0
C
C         READ IN DATA
C THSE STATEMENTS NOW PRECEEDED WITH 'C'S WERE THE ORIGINAL ONES
C THE NEW ONES FOLLOW AND ARE EXPLAINED BY NEW COMMENTS
C 8010 READ(5,7000)NC,NN,NAME
C 8020 IF(NC)8042,8042,8050
C SET CASE NUMBER TO 1-EACH NEW CASE IS INCREMENTED
          NC=1
C TYPE OUT PROGRAM NAME AND ASK FOR NUMBER OF SPINS IN THIS CASE
C
8010      IF(TTFLAG.NE.0)CALL INITT(120)
          TYPE 7100
7100      FORMAT(' LAOCOON3'/' NUMBER OF SPINS= ',$)
9005      ACCEPT 1001, NN
C
C STOP IF NUMBER OF SPINS <2, REASK QUESTION IF >7
C
          IF(NN.GT.7)GO TO 8010
          IF(NN.GT.1)GO TO 9010
C IF ALL DONE FIND OUT IF LPT LISTING IS DESIRED
          TYPE 7275
7275      FORMAT(' PRINT OUTPUT ON LINE PRINTER?(Y OR N): '$)
          ACCEPT7111,AQ
          IF(AQ.NE.'Y')GO TO 7900
C REWIND LPT FILE
          REWIND 24
7850      READ(24,7260,END=7900)BUFFER
7260      FORMAT(27A5)
          WRITE(3,7260)BUFFER
          GO TO 7850
7900      CLOSE(UNIT=24)
          STOP
CC GET TITLE FOR LISTING
C
9010      TYPE 7101
7101      FORMAT(' ENTER TITLE: ',$)
9015      ACCEPT 7102,NAME
7102      FORMAT(15A4)
C 9999 WRITE(LPT,4958)
C
C 8042 STOP
C 8050 READ(5,7002)FR1,FR2,AMIN
C
C GET MINIMUM AND MAXIMUM FREQUENCY OF INTEREST AND MINIMUM INTENSITY
C
9020      TYPE 7110
7110      FORMAT('+MIN. FREQ.= ',$)
9025      ACCEPT 7103, FR1
7103      FORMAT(F)
9030      TYPE 7104
7104      FORMAT('+MAX FREQ.= ',$)
9035      ACCEPT 7103, FR2
C THE MINIMUM INTENSITY IS NOW ENTERED DURING ITERATION ONLY
C
C
C 8060 READ(5,7001)(ISO(I),I=1,7)
C GET LETTERS FOR EACH NUCLEUS IN SYSTEM
C
9050      TYPE 7106
7106      FORMAT(' ENTER LETTERS FOR EACH NUCLEUS: ',$)
9055      ACCEPT 7001,(ISO(I),I=1,7)
C
C 8070 READ(5,7002)(W(I),I=1,7)
C
C GET CHEMICAL SHIFTS
C
9060      DO 9070 I=1,NN
9065      TYPE 7108, I
7108      FORMAT('+V(',I1,')=',$)
```

```
9067      ACCEPT 7103, W(I)
9070      CONTINUE
 8080 NNM=NN-1
 8090 DO 8339J=1,NNM
 8100 JP=J+1
C
C 8110 READ(5,7002)(A(J,K),K=JP,NN)
C
C GET COUPLING CONSTANTS
9075      DO 9080 K=JP,NN
9076      TYPE 7109,J,K
7109      FORMAT('+J(',I1,',',I1,')= ',$)
9077      ACCEPT 7103, A(J,K)
9080      CONTINUE
 8333 DO 8339K=JP,NN
 8334 IF(ISO(J)-ISO(K))8335,8337,8335
 8335 A(K,J)=0.
 8336 GO TO 8339
 8337 A(K,J)=0.5*A(J,K)
 8339 CONTINUE
C
C        CALCULATE SPIN FUNCTIONS IF NOT ALREADY DO NE
C IF NN IS THE SAME AS IN THE PREVIOUS PROBLEM,
C THIS SECTION IS SKIPPED AND THE VALUES IN STORAGE USED.
C
 9350 IF(NN-NNO)9351,160,9351
 9351 NNO=NN
 8351 NO(1)=1
 8352 LL(1)=1
 8353 DO 8357J=1,NN
 8354 JP=J+1
 8355 JD=NN+1-J
 8356 NO(JP)=(NO(J)*JD)/J
 8357 LL(JP)=LL(J)+NO(J)
 8359 NNP=NN+1
 8360 MAX=2**NN
 8361 MAXM=MAX-1
 8362 NP(1)=0
 8363 DO 8371J=1,MAXM
 8364 ISUM=1
 8365 DO 8368M=1,NN
 8366 MEX2=2**M
 8367 IZ=(2*J)/MEX2-2*(J/MEX2)
 8368 ISUM=ISUM+IZ
 8369 K=LL(ISUM)
 8370 NP(K)=J
 8371 LL(ISUM)=LL(ISUM)+1
 8372 DO 8373J=2,NNP
 8373 LL(J)=LL(J)-NO(J)
 8374 DO 8379K=1,MAX
 8375 N=NP(K)
 8376 DO 8379M=1,NN
 8377 MEX2=2**M
 8378 FZ(K,M)=2*(N/MEX2)-(2*N)/MEX2
 8379 FZ(K,M)=FZ(K,M)+0.5
C
C        READ IN ITERATIONS,LINES TO BE MATCHED AND PARAMETERS, IF ANY
C GET NUMBER OF ITERATIONS AND PUNCH PARAMETERS
C IF NKTL(1)=1 A FILE OF THE TRANSITIONS IS CREATED NAMED FOR20.DAT
C   160 READ(5,7001)NI,(NKTL(K),K=1,6)
C
          NI=0
160       TYPE 7113
7113      FORMAT(' ITERATING? (Y OR N): ',$)
9105      ACCEPT 7111,AQ
7111      FORMAT(A1)
9110      IF(AQ.EQ.'Y') GO TO 9120
C
C IF NOT ITERATING FIND OUT IF TRNASITION FILE SHOULD BE CREATED
C
9115      TYPE 7112
7112      FORMAT(' CREATE FILE OF TRANSITIONS? (Y OR N): ',$)
9117      ACCEPT 7111, AQ
```

```
9118    IF(AQ.NE.'Y')GO TO 24
C
C IF A FILE IS TO BE CREATED, GET THE FILE NAME IN 'FNAME'
C IF THE NAME HAS NOW EXTN SPECIFIED, 'DAT' WILL BE USED
C IF THE NAME IS TERMINATED WITH A '.' THERE WILL BE NO EXTN
C
        NKTL(1)=1
        TYPE 7205
7205    FORMAT(' DISK FILE NAME= ',$)
        ACCEPT 7210, FNAME
7210    FORMAT(A10)
C OPEN THAT FILE BY NAME
        OPEN(UNIT=20,FILE=FNAME,ACCESS='SEQOUT')
C NOW GO ON WITHOUT ITERATING
        GO TO 24
C IF ITERATION IS TO BE DO NE, SET MAX ITERATIONS  TO 9 AND PROCEED
C START BY SETTING ITER TO -9, WHICH CAUSES A PASS THROUGH THE
C PROGRAM AS A DIRECT CALCULATION AND THEN RETURNS TO STAEMENT 3
C BELOW TO PRINT OUT THE CALCULATED TRANSITIONS AND GET THE
C ASSIGNMENTS
C
9120    NI=-9
        IF(NI)24,24,3
C GET ASSIGNED TRANSITIONS FROM A FILE ON DISK
C OR ENTER THEM NOW AS DESIRED
C
3       TYPE 7200
7200    FORMAT(' READ IN ASSIGNED TRANS. FROM DISK?(Y OR N): ',$)
        ACCEPT 7111, AQ
        IF(AQ.NE.'Y') GO TO 9200
        TYPE 7205
        ACCEPT 7210,FNAME
C GET MINIMUM INTENSITY
        TYPE 7105
        ACCEPT 7103,AMIN
C OPEN THE INPUT FILE
        OPEN(UNIT=20,FILE=FNAME,ACCESS='SEQIN')
        NL=0
C SKIP OVER CASE NUMBER
        READ(20,1000)ICASE
118 NL=NL+1
C THIS STATEMENT QUITS READING IN IF A ZERO LINE NUMBER IS FOUND
C OR IF THE END OF THE INPUT FILE IS DETECTED
C READ IN LINE # AND ASSIGNED FREQ, SKIPPING CALC FREQ AND INT.
    4 READ(20,1001,END=120)IL(NL),DUM1,DUM2,F(NL)
        IF(IL(NL))120,120,119
C IGNORE 0 FREQUENCIES AS NON-ENTRIES
119     IF(F(NL).EQ.0.)GO TO 4
C ELSE READ ANOTHER ENTRY
        GO TO 118
120 NL=NL-1
        CLOSE(UNIT=20)
C SKIP OVER THE TTY READ IN IF IT WAS READ FROM DISK
        GO TO 121
C GET DATA FROM TTY INSTEAD
9200    TYPE 7105
7105    FORMAT(' MIN. INTENSITY= ',$)
9045    ACCEPT 7103, AMIN
C **********ALL THIS BELOW IS ADDED *********************
C NOW PRINT OUT THE SORTED TRANSITIONS IN THE RANGE FROM
C FR1 TO FR2 AND > AMIN AND GET A NEW VALUE FOR THEM
        NL=1
        DO 9250 I=1,LCT
        IF(DC(I,4).LT.AMIN)GO TO 9250
        IF(DC(I,3).LT.FR1)GO TO 9250
        IF(DC(I,3).GT.FR2)GO TO 9250
C IF THE TRANSITION IS IN RANGE, PRINT OUT THE LINE NUMBER,
C FREQUENCY AND INTENSITY AND GET THE EXPERIMENTAL FREQUENCY
C
        IDC=DC(I,1)+0.5
        TYPE 7215, IDC, DC(I,3),DC(I,4)
7215    FORMAT(1H+,I4,F12.3,F9.3,3X,$)
C
```

```
C GET INPUT VALUE-EITHER RETURN= NO ENTRY
C         NUMBERY=ENTRY
C OR       "        -DUPLICATE PREVIOUS ENTRY
C A DITTO MARK FOR THE FIRST ENTRY IS , OF COURSE, ILLEGAL
C
          ACCEPT 7111, CHAR
          IF(CHAR.EQ.'        ')GO TO 9250
          IF(CHAR.EQ.'"'.AND.NL.EQ.1)GO TO 9250
          IF(CHAR.EQ.'"') GO TO 9240
C REREAD THE LINE AS AN F - FIELD AND STORE THE VALUE
          REREAD 7002, F(NL)
          IL(NL)=IDC
          NL=NL+1
          GO TO 9250
C PUT THE PREVIOUS VALUE IN THIS ONE IF A DITTO MARK WAS TYPED
9240      IL(NL)=IDC
          F(NL)=F(NL-1)
          NL=NL+1
9250      CONTINUE
          NL=NL-1
C
C ****************************************************************
C
C NOW GET THE SETS OF PARAMETERS TO VARY
C IF SINGLE INTEGERS ARE ENTERED, THEY ARE SHIFTS
C IF DOUBLE INTEGERS, THEY ARE COULING CONSTANTS
C
  121 NOS=1
C
C  140 READ(5,1053)(IA(NOS,K),IB(NOS,K),K=1,6)
C
C ENTER PARAMETERS TO BE VARIED. IF SINGLE INTEGERS THEY ARE SHIFTS
C IF DOUBLE INTEGERS, THEY ARE COUPLING CONSTANTS.
C ONLY THOSE ENTERED WILL BE VARIED AT ALL!
C
C ERASE SCREEN HERE IF RUNNING FROM TEKTRONIX
          IF(TTFLAG.NE.0)CALL ERASE
          TYPE 7245
7245      FORMAT(' ENTER PARAMETERS TO BE VARIED:'/)
C PRINT OUT SET NUMBER
9255      TYPE 7246, NOS
7246      FORMAT(1H+,I2,2X,$)
          ACCEPT 7250,(IPAR(K),K=1,6)
7250      FORMAT(6I)
9265      IF(IPAR(1).LE.0) GO TO 150
C CONVERT TO IA AND IB ARRAYS
          DO 9260 I=1,6
C IF COUPLING CONSTANTS, THE LEFT DIGIT GOES IN IA AND THE RIGHT IN IB
          IA(NOS,I)=IPAR(I)/10
          IB(NOS,I)=IPAR(I)-IA(NOS,I)*10
C IF THESE ARE SHIFTS,THE ABOVE IS SPERCEDED: IB IS 0 AND IA IS THE SHIFT #
          IF(IPAR(I).LT.8)IB(NOS,I)=0
          IF(IPAR(I).LT.8)IA(NOS,I)=IPAR(I)
9260      CONTINUE
C LOOP BACK UNTIL A ZERO PARAMETER IS ENTERED
          NOS=NOS+1
          GO TO 9255
  150 NOS=NOS-1
C         ORDER LINES TO BE MATCHED BY ORIGIN
    5 NLM=NL-1
    6 DO   22 J=1,NLM
    7 JP=J+1
    8 ILSM=IL(J)
    9 NOTE=0
   10 DO   14 K=JP,NL
   11 IF(IL(K)-ILSM) 12,14,14
   12 ILSM=IL(K)
   13 NOTE=K
   14 CONTINUE
   15 IF(NOTE) 16,22,16
   16 FS=F(J)
   17 IS=IL(J)
   18 F(J)=F(NOTE)
```

```
   19 IL(J)=ILSM
   20 F(NOTE)=FS
   21 IL(NOTE)=IS
   22 CONTINUE
C FORMERLY KT1=3, KT2=2
   24 KT1=23
   25 KT2=22
C      OUTPUT INITIAL DATA
C THE INPUTDATA IS PRINTED OUT
  535 WRITE(LPT,1002)NC,NAME,NN,FR1,FR2,AMIN
      DO   1535 K=1,NN
      WRITE(LPT,1003)ISO(K),K,W(K)
 1535 CONTINUE
      NNM=NN-1
      DO   1536 J=1,NNM
      JP=J+1
      DO   1536 K=JP,NN
      WRITE(LPT,1004)J,K,A(J,K)
 1536 CONTINUE
 2240 IF(NI)61,61,2250
 2250 WRITE(LPT,1057)
 2260 DO   2330 J=1,NOS
 2261 WRITE(LPT,1058)J
 2263 DO   2320 K=1,6
 2270 IF(IA(J,K))9999,2330,2280
 2280 IF(IB(J,K)) 9999,2290,2310
 2290 WRITE(LPT,1062)IA(J,K)
 2300 GO TO 2320
 2310 WRITE(LPT,1063)IA(J,K),IB(J,K)
 2320 CONTINUE
 2330 CONTINUE
C      ENTER ITERATIVE LOOP
C INITIALIZATION OF PARAMETERS; READY THE SCRATCH FILES
   61 ER1=10000.0
   62 ITER=0
C FORMERLY REWIND 2, REWIND 3
  400 REWIND 22
  401 REWIND 23
C
C
C NA IS THE NUMBER OF THE SUBMATRIX OF THE HAMILTONIAN
C ABOUT TO BE EVALUATED. MOA IS THE ORDER OF THAT MATRIX
C
  402 NA=1
  403 MOA=NO(NA)
C
C      COMPUTE ELEMENTS OF HAMILTONIAN MATRIX
C THE SUBMATRIX IS EVALUATED. J AND K ARE ROW AND
C COLUMNS OF THE SUBMATRIX, JL AND KL ARE THE SUBSCRIPTS
C OF THE CORRESPONDING BASIS FUNCTIONS (NP) AND
C Z-COMPONENTS (FZ). IF J=K A DIAGONAL ELEMENT IS BEING COMPUTED
C AND STATEMENTS 425-432 OPERATE.
C IF J =/ K, AN OFF-DIAGONAL ELEMENT IS BEING COPUTED AND
C STATEMENTS 408-424 OPERATE.
C IF NO IERATION HAS BEEN PERFORMED, THE VARIABLE KNTRL IS SET
C TO ZERO AND EXECUTION SKIPS TO STATEMENT 512. IF AN
C ITERATION HAS BEEN PERFORMED, THE NEXT EXECUTED STATEMENT IS 500.
C
 1403 DO   432 J=1,MOA
  404 DO   432 K=J,MOA
  405 JL=LL(NA)+J-1
  406 KL=LL(NA)+K-1
  407 IF(J-K)408,425,408
  408 KINV=0
  409 DO   418 M=1,NN
  410 P=FZ(JL,M)*FZ(KL,M)
  411 IF(P)412,418,418
  412 KINV=KINV+1
  413 IF(KINV-1)418,414,416
  414 MA=M
  415 GO TO 418
  416 IF(KINV-2)418,417,422
  417 MB=M
  418 CONTINUE
```

```
  419 X(J,K)=A(MB,MA)
  420 X(K,J)=X(J,K)
  421 GO TO 432
  422 X(J,K)=0.
  423 X(K,J)=0.
  424 GO TO 432
  425 X(J,J)=0.
  426 DO 427 M=1,NN
  427 X(J,J)=X(J,J)+FZ(JL,M)*W(M)
  428 DO   431 M=1,NNM
  429 MP=M+1
  430 DO   431 N=MP,NN
  431 X(J,J)=X(J,J)+FZ(JL,M)*FZ(JL,N)*A(M,N)
  432 CONTINUE
      IF(ITER)8042,1500,500
C
C THESE WERE MOVED HERE FROM AFTER STATEMENT 9015
9999    WRITE(LPT,4958)
8042    STOP
C
 1500 KNTRL=0
      GO TO 512
C
C        ROUGH DIAGONALIZATION
C EIGENVECTORS WRITTEN ON TAPE KT2 DURING THE PREVIOUS ITERATION
C ARE READ OFF AND ARE USED TO PERFORM A ROUGH DIAGONALIZATION OF
C THE SUBMATRIX.  KNTRL IS SET TO 1.
C
  500 READ(KT2)((VB(J,K),J=1,MOA),K=1,MOA)
  501 DO   505 JA1=1,MOA
  502 DO   505 JA2=1,MOA
  503 VA(JA1,JA2)=0.
  504 DO   505 JX1=1,MOA
  505 VA(JA1,JA2)=VA(JA1,JA2)+X(JA1,JX1)*VB(JX1,JA2)
  506 DO   510 JX1=1,MOA
  507 DO   510 JX2=1,MOA
  508 X(JX1,JX2)=0.
  509 DO   510 JA1=1,MOA
  510 X(JX1,JX2)=X(JX1,JX2)+VB(JA1,JX1)*VA(JA1,JX2)
C        FINISH DIAGONALIZATION
C
  511 KNTRL=1
C IF KNTRL=0 MATRIX DIAGONALIZES FROM SCRATCH.
C
C
C THE VARIABLE UNIT # LPT IS ADDED TO THE CALL TO MATRIC
C SINE IT DOES NOT HAVE A COMMMON STATEMENT
C
  512 CALL MATRIX(X,VB,MOA,KNTRL,LPT)
C
C        STORE ENERGIES AND EIGENVECTORS
C
  513 DO   515 J=1,MOA
  514 JL=J+LL(NA)-1
C
C EIGENVECTORS ARE STORED
C
  515 E(JL)=X(J,J)
C
C EIGENVECTORS ARE STORED ON TAPE KT1
C
 1514 WRITE(KT1)((VB(J,K),J=1,MOA),K=1,MOA)
      IF(NI) 516,516,1515
C
C CALLS SUBROUTINE DIFFER IF ITERATIONS ARE TO BE PERFORMED
 1515 CALL DIFFER (MOA,NA)
C
C NA IS INCREASED AND TESTED TO SEE THAT ALL SUBMATRICES HAVE
C BEEN PROCESSED. IN NOT, GO BACK TO 403
C IF DONE GO ON TO 1518. IF NO ITERATIONS, GO ON TO 602
C
  516 NA=NA+1
  517 IF(NA-NN-1) 403,403,1518
 1518 IF(NI) 602,602,518
```

```
C
C       LEAST SQUARES ROUTINES
C THE EQUATIONS OF CONDITION ARE FORMED
C
  518 CALL CONDIT
C
C THE RMS ERROR IS EVALUATED AND DECISION MADE WHETHER TO
C ITERATE FURTHER.
C
  519 CALL ERROR9(NEXIT,ER1,ITER)
C
C THE NORMAL EQNS OF THE LEAST-SQUARES METHOD ARE FORMED
C
  521 CALL NORMAL
C
C IF NO MORE ITERATIONS ARE TO BE PERFORMED, CONTROL PASSES TO 523
C IF MORE THEN TO 526
C
  522 IF(NEXIT) 9999,523,526
C
C THE MATRIX OF NORMAL EQUATIONS IS INVERTED
C
  526 CALL INVERT(VA,NOS)
C
C THE CORRECTIONS (CORR) TO BE APPLIED TO THE PARAMETERS SETS
C ARE EVALUATED.
C
  527 DO   530 NSA=1,NOS
  528 CORR(NSA)=0.
  529 DO   530 NSB=1,NOS
  530 CORR(NSA)=CORR(NSA)+VA(NSA,NSB)*BV(NSB)
C
C       APPLY COMPUTED CORRECTIONS
C
  531 CALL CORREC
C
C THE IDENTITIES OF TAPES KT1 AND KT2 ARE SWITCHED.
C THE EIGENVECTORS WRITTEN ON KT1 ARE NOW AVAILABLE ON KT2
C FOR THE NEXT CYCLE OF ITERATION.
C
  532 KT1=KT2
C
C  533 KT2=6/KT2
C CONVERT BY DIVIDING INTO 6, BUT REMOVE 20 BIAS FIRST
C
533      KT2=(6/(KT2-20))+20
C
C THE ITERATIONCOUNTER IS INCREASED AND A NEW
C CYCLE OF ITERATION IS STARTED AT STATEMENT 400
C
  520 ITER=ITER+1
  534 GO TO 400
C
C       END OF LOOP. OUTPUT REFINED PARAMETERS
C CONTROL WAS TRANSFERRED FROM 522, ITERATIONS HAVING BEEIN PERFORMED
C AND NOW TERMINATED. THE ADJUSTED PARAMETERS ARE PRINTED OUT.
C
  523 KNTRL=0
C PRINT MESSAGE ON TTY-FINAL VALUES
C ERASE SCREEN IF TEKTRONIX
         IF(TTFLAG.NE.0)CALL ERASE
         TYPE 7400, ITER
7400     FORMAT(' FINAL VALUES, ITERATION ',I3)
         WRITE(LPT,1012)NC
  536 DO   538 K=1,NN
  537 WRITE(LPT,1003)ISO(K),K,W(K)
C PRINT FINAL SHIFTS ON TTY
         TYPE 7405, K,W(K)
7405     FORMAT(' V(',I1,')=',F10.3)
  538 CONTINUE
  539 DO   543 K=1,NNM
  540 KP=K+1
  541 DO   543 L=KP,NN
  542 WRITE(LPT,1004)K,L,A(K,L)
```

```
C TPYE FINAL J'S ON TTY
         TYPE 7440, K,L,A(K,L)
7440     FORMAT(' J(',I1,',',I1,')=',F10.3)
   543 CONTINUE
   544 WRITE(LPT,1005)
C        BEGIN ERROR ANALYSIS
C THE MATRIX OF NORMAL EQNS IS DAGONALIZED
C
   524 CALL MATRIX(VA,VB,NOS,KNTRL,LPT)
C
C THE STANDARD ERROR VECTORS AND PROBABLY ERRORS (CORR)
C ARE COMPUTED AND PRINTED OUT. IF NO ITERATION HAS BEEN PERFRMED,
C THE VARIABLE KNTRL IS SET TO 0 AND EXECUTION  GOES TO 512.
C IF AN ITERATION HAS BEEN PERFORMED, THE NEXT STATEMENT IS 500.
C
   635 FNL=NL
   636 FNS=NOS
   637 DEV=(ER1*ER1*FNL)/(FNL-FNS)
   638 DO  642 NS=1,NOS
   639 ER2=SQRT(DEV/VA(NS,NS))
C WRITE ERROR VECTORS
   640 WRITE(LPT,1006)(VB(K,NS),K=1,NOS)
C WRITE STANDARD ERRORS
   641 WRITE(LPT,1007)ER2
  1641 DO  1642 J=1,NOS
  1642 CORR(J)=CORR(J)+(VB(J,NS)*ER2)**2
   642 CONTINUE
C PRINT MESSAGE ON TTY
         TYPE 7415
7415     FORMAT(' ERRORS IN EACH PARAMETER SET')
   700 DO 701NS=1,NOS
   701 CORR(NS)=0.6745*SQRT(CORR(NS))
C WRITE OUT ERRORS IN PARAMETERS SETS
         TYPE 7420,(NS,CORR(NS),NS=1,NOS)
7420     FORMAT(I3,2X,F10.3)
   702 WRITE(LPT,7003)(NS,CORR(NS),NS=1,NOS)
C        CALCULATE INTENSITIES AND OUTPUT LINES
   601 LM=1
  1601 IZ=1
   602 LINE=0
       LCT=0
       NOTA=1
C
C INITIALIZZE COUNTERS AND PREPARE TO COMPUTE INTENSITIES
C AND PRINT RESULTS. IN DIRECT CALCULATION, CONTROL PASSED HERE
C FROM 1518.
C
C  600 WRITE(LPT,1009)NC
  600       WRITE(LPT,1009)NC
   603 REWIND KT1
  1603 READ(KT1)VB(1,1)
   604 NA=0
C
C NA AND NB ARE SUBMATRICES BETWEEN WHICH TRANSITIONS WILL BE COMPUTED
C MOA AND MOB ARE THE CORRESPONDING MATRIX ORDERS.
C
   605 NA=NA+1
   606 NB=NA+1
   607 MOA=NO(NA)
   608 MOB=NO(NB)
C
C THE MATRIX PRODUCT OF SA, THE EIGENVECTORS OF SUBMATRIX NA
C AND IX IS FORMED AND STORED IN VA.
C
   609 DO  624 JB=1,MOB
  1609 DO  1610 JA=1,MOA
  1610 VA(JA,JB)=0.
   610 NPB=JB+LL(NB)-1
   611 DO  624 JA=1,MOA
   612 NPA=JA+LL(NA)-1
   613 NDIF=NP(NPB)-NP(NPA)
   614 DO  617 M=1,NN
   615 MEX2=(2**M)/2
   616 IF(NDIF-MEX2) 617,622,617
```

```
      617 CONTINUE
      619 GO TO 624
      622 DO    623 JC=1,MOA
      623 VA(JC,JB)=VA(JC,JB)+VB(JA,JC)
      624 CONTINUE
C
C SB, THE EIGENVECTORS OF SUBMATRIX NB ARE RETRIEVED FROM TAPE.
C
      625 READ(KT1)((VB(J,K),J=1,MOB),K=1,MOB)
C
C THE INTENSITY (S) OF THE TRANSITION FROM KA IN MATRIX NA
C TO KB IN MATRIC NB IS COMPUTED.
      658 DO    671 KA=1,MOA
      659 DO    671 KB=1,MOB
      660 LINE=LINE+1
      663 S=0.
      664 DO    665 KD=1,MOB
      665 S=S+VA(KA,KD)*VB(KD,KB)
      666 S=S*S
C
C THE FREQUENCY (FR) OF THE SAME TRANSITION IS FOUND.
C
     1666 KAL=KA+LL(NA)-1
      667 KBL=KB+LL(NB)-1
      668 FR=E(KAL)-E(KBL)
          IF(NI)1669,1669,1668
C RESULTS ARE PRINTED
     1668 IF(LINE-IL(LM))1669,669,1669
     1669 IF(S-AMIN) 671,2669,2669
     2669 IF(FR-FR1) 671,4669,4669
     4669 IF(FR2-FR) 671,1670,1670
C WRITE OUT LINES AND INTENSITIES ONLY IF NOT PREPARING TOITERATE
     1670    IF(NI.GE.0)WRITE(LPT,1016)LINE,FR,S
          LCT = LCT+1
          IF(LCT-300) 3670,3670,3671
3670      DC(LCT,1)=LINE
          DC(LCT,2)=0.
          DC(LCT,3)=FR
          DC(LCT,4)=S
          GO TO 671
     3671 NOTA=0
     1671 GO TO 671
      669 WRITE(LPT,1010)LINE,F(LM),FR,S,B(LM)
          LCT = LCT+1
          IF(LCT-300) 3669,3669,670
3669      DC(LCT,1)=LINE
          DC(LCT,2)=F(LM)
          DC(LCT,3)=FR
          DC(LCT,4)=S
          DC(LCT,5)=B(LM)
      670 LM=LM+1
      671 CONTINUE
C NA IS TESTED TO SEE IF AL SUBMATRICES HAVE BEEN TREATED.
C IF NOT, CONTROL IS RETURNED TO STATEMENT 605..
C IF YES, CONTROL PROGRESSES TO 930.
      672 IF(NN-NA)930,930,605
C          ORDER LINES IN STORAGE
C THE LINE NUMBERS, EXPERIMENTAL AND CALCULATED FREQUENCIES,
C INTENSITIES AND ERRORS, ARE REARRANGED IN ORDER OF INCREASING
C CALCULATED FREQUENCY
      930 LCT=AMINO(LCT,300)
          LCTM=LCT-1
      931 DO    948 J=1,LCTM
     1931 NOTE=0
      932 FSM=DC(J,3)
          JP=J+1
      933 DO    937 K=JP,LCT
      934 IF(FSM-DC(K,3)) 937,937,935
      935 FSM=DC(K,3)
      936 NOTE=K
      937 CONTINUE
      938 IF(NOTE) 9999,948,939
      939 DO    940 KT=1,5
```

```
      940 FNO(KT)=DC(J,KT)
      941 DO   942 KT=1,5
      942 DC(J,KT)=DC(NOTE,KT)
      943 DO   944 KT=1,5
      944 DC(NOTE,KT)=FNO(KT)
      948 CONTINUE
          IF(NKTL(1))1950,1950,949
C       OUTPUT ORDERED LINES
C "PUNCH" HEADER CARD IF PUNCHED OUTPUTREQUIRED
      949 WRITE(20,1000)NC
C PRINT 2ND TABLE , PUNCH FREQUEICIES AND INTENISITIES OF REQD
1950      WRITE(LPT,4105) NC
      950 DO   955 L=1,LCT
          IDC = DC(L,1)+0.5
          IF(NKTL(1))960,960,965
      965 WRITE(20,7004)IDC,DC(L,3),DC(L,4)
C WRITE OUT LISTING IN ORDER ONLY IF ITER>=0
      960 IF(NI)956,952,951
      951 IF(DC(L,2)) 954,952,954
      952 WRITE(LPT,1016)IDC,DC(L,3),DC(L,4)
      953 GO TO 955
      954 WRITE(LPT,1010)IDC,(DC(L,J),J=2,5)
      955 CONTINUE
C   956 IF(NOTA)673,957,673
956       IF(NOTA)9270,957,9270
      957 WRITE(LPT,4957)
C NOW GO BACK AND ITERATE IF NI IS NEGATIVE
C IF 0 THIS WAS DIRECT AND CASE IS DONE
C IF WE WERE WRITING TRANSITONS ONTO DISK, CLOSE THE FILE
9270      IF(NKTL(1).EQ.1)CLOSE(UNIT=20)
          IF(NI)9300,673,673
C
C NEGATE NI AND START ITERATING
9300      NI=-NI
          GO TO 3
      673 WRITE(LPT,1011)NC
C       GO ON TO NEXT PROBLEM
C INCREMENT CASE NUMBER BY 1
          NC=NC+1
          WL=0.5
          Q=(2/WL)**2
C SKIP ALL THIS IF NO PLOT POSSIBLE
          IF(TTFLAG.EQ.0)GO TO 8010
          TYPE 790
790       FORMAT(' DO YOU WANT A PLOT OF THESE DATA?(Y OR N): '$)
          ACCEPT 7007, ANS
          IF(ANS.NE.'Y')GO TO 8010
800       DO 801 I=1,1024
801       SUM(I)=0
C GENERATE LORENTZIAN ARRAY
          HZPPT=(FR2-FR1)/1024.
          DO 820 I=1,LCT
C JD IS J INCREMENT ONEITHER SIDE OF CENTER
C DELTAF IS CORRESPONDING FREQUENCY INCREMENT
          JD=0
          DELTAF=0
          IF(DC(I,3).LT.FR1)GO TO 820
          IF(DC(I,3).GT.FR2)GO TO 820
C J IS INDEX OF CENTER OF POINT. LOW FREQ END AT LEFT
          J=1024-(DC(I,3)-FR1)/HZPPT
          GMAX=DC(I,4)
810       GF=GMAX/(1.+Q*DELTAF**2)
811       IF(GF.LT..001)GO TO 820
          IF(((J-JD).LT.1).AND.((J+JD).GT.1024))GO TO 820
          IF((J-JD).GE.1)SUM(J-JD)=SUM(J-JD)+GF
C SKIP A 2ND ADDITION IF THIS IS THE CENTER POINT
          IF(JD.EQ.0)GO TO 812
          IF((J+JD).LE.1024)SUM(J+JD)=SUM(J+JD)+GF
812       DELTAF=HZPPT+DELTAF
          JD=JD+1
          GO TO 810
820       CONTINUE
825       CALL ERASE
```

```
C PRINT TITLE IF THIS IS TEKTRONIX
        CALL MOVABS(1,750)
        CALL ANMODE
              TYPE 7005,NAME
        CALL MOVABS(1,100)
C SCALE TO 600 MAX-FIRST FIND CURRENT MAX
        FMAX=0
        DO 828 I=1,1024
        IF(SUM(I).GT.FMAX)FMAX=SUM(I)
828     CONTINUE
        DO 829 I=1,1024
829     SUM(I)=SUM(I)*500./FMAX +100
        DO 830 I=1,1024
        IY=SUM(I)
        K=I-1
830     CALL DRWABS(K,IY)
        CALL MOVABS(1,90)
        CALL ANMODE
        TYPE 7008,FR2,FR1
7008     FORMAT(1X,F8.2,55X,F8.2)
        TYPE 840
840     FORMAT(' PLOT NEW FREQ RANGE? (Y OR N): '$)
        ACCEPT 7007,ANS
7007    FORMAT(A1)
        IF(ANS.NE.'Y')GO TO 8010
        TYPE 7110
        ACCEPT 7103,FR1
        TYPE 7104
        ACCEPT 7103, FR2
C MAKE SURE THAT FR1<FR2
        IF(FR2.GT.FR1)GO TO 800
C OTHERWISE SWITCH THEM
        T=FR1
        FR1=FR2
        FR2=T
        GO TO 800
7005    FORMAT(1X,15A4)
1000 FORMAT(I3)
1001 FORMAT(I,3F)
1002 FORMAT(1H1,5X,'LAOCOON III'1X,4HCASE,I4,5X,15A4///21X,3HN
     1N=,I2,17H   FREQUENCY RANGE,2F10.3,20H   MINIMUM INTENSITY ,F7.5///2
     21X,16HINPUT PARAMETERS//)
C 1003 FORMAT(1H ,20X,I1,9X,2HW(I1,2H)=F9.3)
1003 FORMAT(1H ,20X,A1,9X,2HV(I1,2H)=F9.3)
C 1004 FORMAT(1H ,30X,2HJ(I1,1H,I1,2H)=F9.3)
1004 FORMAT(1H ,30X,2HA(I1,1H,I1,2H)=F9.3)
1005 FORMAT(1H0,19X,37HERROR VECTORS AND STANDARD ERRORS
1006 FORMAT(1H ,20X,12F8.4)
1007 FORMAT(1H ,30X,15HSTANDARD ERROR=F8.3)
1009 FORMAT(1H1,19X,8HCASE NO I4//20X,'LINE       EXP FREQ   CALC FREQ'
     1,'    INTEN   ERROR' //)
1010 FORMAT(1H ,19X,I4,F11.3,F12.3,F9.3,F12.3)
1011 FORMAT(1H0,20X,12HEND OF CASE I3)
1012 FORMAT(1H1,20X,16HBEST VALUES CASE I4//)
1015 FORMAT(10F12.4)
1016 FORMAT(1H ,19X,I4,F23.3,F9.3)
1053 FORMAT(6(I1,I1,2X))
1057 FORMAT(1H0,24X,14HPARAMETER SETS)
C 1062 FORMAT(1H ,24X,2HW(I1,1H))
1062 FORMAT(1H ,24X,2HV(I1,1H))
C 1063 FORMAT(1H ,24X,2HA(I1,1H,I1,1H))
1063 FORMAT(1H ,24X,2HJ(I1,1H,I1,1H))
1058 FORMAT(1H ,I21)
4105 FORMAT(1H1,20X,19HORDERED LINES CASE I4/// 21X,4HLINE,3X,8HEXP FRE
     1Q,3X,9HCALC FREQ,3X,5HINTEN,3X,5HERROR///)
4957 FORMAT(1H0,20X,15HDC STORAGE FULL)
4958 FORMAT(1H ,25X,16H IMPOSSIBLE DATA)
C FORMAT CHANGED TO A4
7000 FORMAT(2I,10A4)
7001 FORMAT(7A1)
C FORMAT CHANGED FROM 7F10.3 TO 7F
7002 FORMAT(7F)
```

```
7003 FORMAT(1H0,19X,33HPROBABLE ERRORS OF PARAMETER SETS //(1H ,20X,
    1I2,F10.3))
7004    FORMAT(I5,2X,2F)
     END




     SUBROUTINE MATRIX(X,E,N,KNTRL,LPT)
C DIAGONALIZES A REAL SYMMETRIC MATRIX BY THE JACOBI METHOD
     DIMENSION X(35,35),E(35,35)
     NSQP=3*N*N/2
     ITER=0
     NMI=N-1
     IF(NMI)87,2,3
   2 E(1,1)=1.
     GO TO 87
   3 IF(KNTRL) 7,9,7
C    ENTER E MATRIX
   9 DO 11I=1,N
     DO 10J=1,N
  10 E(I,J)=0.0
  11 CONTINUE
     DO 12K=1,N
  12 E(K,K)=1.
C    FIND LARGEST OFF-DIAGONAL ELEMENT
   7 ITER=ITER+1
  21 BIGX=0.0
  22 DO 25J=2,N
     JM1=J-1
     DO 25I=1,JM1
     IF(BIGX-ABS(X(I,J)))23,25,25
  23 BIGX=ABS(X(I,J))
     K=I
     L=J
  25 CONTINUE
     IF (BIGX-0.00001) 76,76,27
C    COMPUTE ROTATION
  27 TS=X(K,L)*X(K,L)
  28 DEL=X(K,K)-X(L,L)
     R=SQRT(ABS(DEL*DEL+4.*TS))
     A=SQRT(ABS((R+DEL)/(2.*R)))
     IF(.707-A)30,30,29
  29 B=-A
     A=SQRT(1.-B*B)
     GO TO 32
  30 B=-SQRT(1.-A*A)
  32 IF(DEL/X(K,L))33,60,60
  33 B=-B
C    ENTER ROTATION IN EIGENVECTOR MATRIX
  60 DO 65J=1,N
     F1=E(J,K)
     E(J,K) = A*F1-B*E(J,L)
  75 E(J,L) = B*F1+A*E(J,L)
C       ORTHOGONAL ROTATION OF MATRIX
     IF(K-J)61,65,61
  61 IF(L-J)62,65,62
  62 X(K,J)=A*X(J,K)-B*X(J,L)
     X(L,J)=B*X(J,K)+A*X(J,L)
  67 X(J,K)=X(K,J)
     X(J,L)=X(L,J)
  65 CONTINUE
     D=X(K,K)+X(L,L)
     X(K,K)=A*A*X(K,K)+B*B*X(L,L)-2.*A*B*X(K,L)
     X(L,L)=D-X(K,K)
     X(L,K)=0.0
     X(K,L)=0.0
     IF(ITER-NSQP)7,90,90
  90 WRITE(LPT,91)
  91 FORMAT(26H TOUGH MATRIX  SKIPPED OUT)
  87 RETURN
  76 BIGR=0.00004
```

```
      DO 86I=1,NMI
      IPI=I+1
      DO 86J=IPI,N
      IF(X(I,J))77,86,77
   77 DEL=X(I,I)-X(J,J)
      IF(DEL) 79,86,79
   79 ROT=ABS(X(I,J)/DEL)
      IF(ROT-BIGR)86,86,78
   78 BIGR=ROT
      K=I
      L=J
   86 CONTINUE
      IF(BIGR-0.00005) 87,87,88
   88 WRITE(LPT,89)
   89 FORMAT(24H LARGE RESIDUAL ROTATION)
      GO TO 27
      END



      SUBROUTINE INVERT(A,N)
C THIS SUBROUTINE INVERTS A LARGE SYMMETRIC POSITIVE MATRIX
C TAKEN DIRECTLY FROM 'A COMPARISON OF SEVERAL METHODS...'
C BY M.H. LIETZE ET. AL., ORNL-3430 (9-MAY-1963)
      DIMENSION A(35,35),B(35),C(35),LZ(35)
      DO  10 J=1,N
   10 LZ(J)=J
      DO  20 I=1,N
      K=I
      Y=A(I,I)
      L=I-1
      LP=I+1
      IF(N-LP)14,11,11
   11 DO 13 J=LP,N
      W=A(I,J)
      IF(ABS(W)-ABS(Y))13,13,12
   12 K=J
      Y=W
   13 CONTINUE
   14 DO 15J=1,N
      C(J)=A(J,K)
      A(J,K)=A(J,I)
      A(J,I)=-C(J)/Y
      A(I,J)=A(I,J)/Y
   15 B(J)=A(I,J)
      A(I,I)=1.0/Y
      J=LZ(I)
      LZ(I)=LZ(K)
      LZ(K)=J
      DO 19K=1,N
      IF(I-K)16,19,16
   16 DO 18J=1,N
      IF(I-J)17,18,17
   17 A(K,J)=A(K,J)-B(J)*C(K)
   18 CONTINUE
   19 CONTINUE
   20 CONTINUE
      DO 200I=1,N
      IF(I-LZ(I))100,200,100
  100 K=I+1
      DO 500J=K,N
      IF(I-LZ(J))500,600,500
  600 M=LZ(I)
      LZ(I)=LZ(J)
      LZ(J)=M
      DO 700L=1,N
      C(L)=A(I,L)
      A(I,L)=A(J,L)
  700 A(J,L)=C(L)
  500 CONTINUE
  200 CONTINUE
      RETURN
      END
```

```
      SUBROUTINE CORREC
C THIS SUBROUTINE APPLIES THE COMPUTED CORRECTIONS TO THE
C PARAMETERS BEING VARIED.
      DIMENSION IL(300),F(300),NAME(15),ISO(7),W(7),A(7,7),IA(29,6),IB(2
     19,6),FZ(128,7),NO(8),LL(8),NP(128),B(300),DC(300,28),X(35,35),D(12
     28,28),VA(35,35),BV(28),VB(35,35),CORR(28),E(128)
      COMMON NOS,NN,NL,NI,IL,F,NAME,ISO,W,A,IA,IB,FZ,NO,LL,NP,B,DC,X,D,V
     2A,BV,VB,CORR,E,LPT
  300 DO  310 NS=1,NOS
  301 DO  309 K=1,6
  302 IF(IA(NS,K)) 303,310,303
  303 IAS=IA(NS,K)
 1303 IF(IB(NS,K)) 304,308,304
  304 IBS=IB(NS,K)
 1304 A(IAS,IBS)=A(IAS,IBS)+CORR(NS)
  305 IF(ISO(IAS)-ISO(IBS)) 309,306,309
  306 A(IBS,IAS)=A(IAS,IBS)/2.
  307 GO TO 309
  308 W(IAS)=W(IAS)+CORR(NS)
  309 CONTINUE
  310 CORR(NS)=0.0
  311 RETURN
      END




      SUBROUTINE ERROR9(NEXIT,ER1,ITER)
C THIS SUBROUTINE CONTROLS WHETHER FURTHER ITERATIONS ARE TO BE MADE
C BY EVALUATING THE RMS ERROR IN FITTING ER2, DETERMINGIN WHETHER
C IT HAS BEEN REDUCED BY MORE THAN 1% FROM THE PREVOUOS VALUE ER1,
C AND WHETHER THE MAX NUMBER OF ITERATIONS)NI HAVE BEEN PERFOMED
C AND RETRUNIN THE CONTROL VARIABLE NEXIT THE VALUE 0 IF NO
C MORE ITERATIONS ARE TO BE PERFORMED AND 1 IF MROE ARE TO BE DONE.
      DIMENSION IL(300),F(300),NAME(15),ISO(7),W(7),A(7,7),IA(29,6),IB(2
     19,6),FZ(128,7),NO(8),LL(8),NP(128),B(300),DC(300,28),X(35,35),D(12
     28,28),VA(35,35),BV(28),VB(35,35),CORR(28),E(128)
      COMMON NOS,NN,NL,NI,IL,F,NAME,ISO,W,A,IA,IB,FZ,NO,LL,NP,B,DC,X,D,V
     2A,BV,VB,CORR,E,LPT
    1 ER2=0.0
    2 FNL=NL
    3 DO  4 K=1,NL
    4 ER2=ER2+B(K)*B(K)
    5 ER2=SQRT(ER2/FNL)
    6 WRITE(LPT,3000)ITER,ER2
    7 IF((ER1-ER2)/ER1-0.01) 8,8,10
    8 NEXIT=0
   18 ER1=ER2
    9 RETURN
   10 IF(ITER-NI) 110,8,8
  110 ER1=ER2
   11 NEXIT=1
   12 RETURN
 3000 FORMAT(1H ,20X,9HITERATIONI3,'     R M S ERROR = 'F8.3)
      END




      SUBROUTINE DIFFER(MOA,NA)
C FORMS THE MATRIX DYI/DPJ, THE VALUES BEING STORED IN D(JL,NS)
C NA AND MOA ARE THE CURRENT NUMBER AND ORDER OF THE SUBMATRIX
      DIMENSION IL(300),F(300),NAME(15),ISO(7),W(7),A(7,7),IA(29,6),IB(2
     19,6),FZ(128,7),NO(8),LL(8),NP(128),B(300),DC(300,28),X(35,35),D(12
     28,28),VA(35,35),BV(28),VB(35,35),CORR(28),E(128)
      COMMON NOS,NN,NL,NI,IL,F,NAME,ISO,W,A,IA,IB,FZ,NO,LL,NP,B,DC,X,D,V
     2A,BV,VB,CORR,E,LPT
   31 DO  58 J=1,MOA
   32 JL=J+LL(NA)-1
   33 DO  58 NS=1,NOS
   34 D(JL,NS)=0.
```

```
 35 DO   57 NPS=1,6
 36 IF(IA(NS,NPS)) 9999,58,37
 37 IAS=IA(NS,NPS)
 38 IF(IB(NS,NPS)) 9999,39,43
 39 DO   41 K=1,MOA
 40 KL=K+LL(NA)-1
 41 D(JL,NS)=D(JL,NS)+VB(K,J)*VB(K,J)*FZ(KL,IAS)
 42 GO TO 57
 43 IBS=IB(NS,NPS)
 44 DO   46 K=1,MOA
 45 KL=K+LL(NA)-1
 46 D(JL,NS)=D(JL,NS)+VB(K,J)*VB(K,J)*FZ(KL,IAS)*FZ(KL,IBS)
 47 IF(ISO(IAS)-ISO(IBS)) 57,48,57
 48 MOAM=MOA-1
148 IF(MOAM)57,57,54
 54 IDIF=(2**IBS-2**IAS)/2
 49 DO 157 JA=1,MOAM
 50 JAL=JA+LL(NA)-1
 51 JAP=JA+1
 52 DO 157 KA=JAP,MOA
 53 KAL=KA+LL(NA)-1
 55 IF(NP(KAL)-NP(JAL)-IDIF)157,66,157
 66 SS=0.
 67 DO 68KM=1,NN
 68 SS=SS+(FZ(KAL,KM)-FZ(JAL,KM))**2
 69 NTST=SS+0.1
 70 IF(NTST-2)157,56,157
 56 D(JL,NS)=D(JL,NS)+VB(JA,J)*VB(KA,J)
157 CONTINUE
 57 CONTINUE
 58 CONTINUE
 59 RETURN
9999 STOP
    END



    SUBROUTINE NORMAL
C COMPILES THE COEFFICIENTS OF THE NORMAL EQUATIONS VA,
C FROM THE COEFFICIENTS OF THE EQUATIONS,DC. THE CONSTANT TERMS,BV,
C ARE LIKEWISE CALCULATED.
    DIMENSION IL(300),F(300),NAME(15),ISO(7),W(7),A(7,7),IA(29,6),IB(2
   19,6),FZ(128,7),NO(8),LL(8),NP(128),B(300),DC(300,28),X(35,35),D(12
   28,28),VA(35,35),BV(28),VB(35,35),CORR(28),E(128)
    COMMON NOS,NN,NL,NI,IL,F,NAME,ISO,W,A,IA,IB,FZ,NO,LL,NP,B,DC,X,D,V
   2A,BV,VB,CORR,E,LPT
207 DO   210 NS1=1,NOS
202 DO   206 NS2=NS1,NOS
203 VA(NS1,NS2)=0.
204 DO   205 LEQ=1,NL
205 VA(NS1,NS2)=VA(NS1,NS2)+DC(LEQ,NS1)*DC(LEQ,NS2)
206 VA(NS2,NS1)=VA(NS1,NS2)
208 BV(NS1)=0.
209 DO   210 LEQ=1,NL
210 BV(NS1)=BV(NS1)+DC(LEQ,NS1)*B(LEQ)
211 RETURN
    END



    SUBROUTINE CONDIT
C THIS SUBROUTINE COMPILES THE COEFFICIENTS OF THE EQNS OF CONDITION.
    DIMENSION IL(300),F(300),NAME(15),ISO(7),W(7),A(7,7),IA(29,6),IB(2
   19,6),FZ(128,7),NO(8),LL(8),NP(128),B(300),DC(300,28),X(35,35),D(12
   28,28),VA(35,35),BV(28),VB(35,35),CORR(28),E(128)
    COMMON NOS,NN,NL,NI,IL,F,NAME,ISO,W,A,IA,IB,FZ,NO,LL,NP,B,DC,X,D,V
   2A,BV,VB,CORR,E,LPT
 99 LUL=0
100 NA=0
101 K=1
102 NA=NA+1
```

```
103 NB=NA+1
104 IF(NA-NN) 105,105,120
105 LLL=LUL+1
106 MOA=NO(NA)
107 MOB=NO(NB)
108 LUL=LUL+MOA*MOB
109 IF(LUL-IL(K)) 102,110,110
110 IDX=IL(K)-LLL
111 JA=(IDX/MOB)+1
112 JB=IDX-(JA-1)*MOB +1
113 JAL=JA+LL(NA)-1
114 JBL=JB+LL(NB)-1
115 B(K)=F(K)+E(JBL)-E(JAL)
116 DO   117 NS=1,NOS
117 DC(K,NS)=D(JAL,NS)-D(JBL,NS)
118 K=K+1
119 IF(K-NL) 109,109,120
120 RETURN
    END
```

```
C THE FOLLOWING DUMMY ROUTINES ARE TO BE COMPILED WHEN
C THE SYSTEM IN USE IS NOT A DEC-10 SUPPORTING A TEKTRONIX GRAPHICS TERMINAL
C THE ASSEMBLY LANGUAGE ROUTINES FOR CONTROLLING
C THIS DISPLAY TERMINAL ARE AVAILABLE FROM THE MANUFACTURER
C AND ARE CALLED FROM WITHIN THE FORTRAN LAOCOON III PROGRAM.
C
C TO COMPILE THESE DUMMY ROUTINES,REMOVE THE 'C' S.
C
C         SUBROUTINE ERASE
C         RETURN
C         END
C
C         SUBROUTINE INITT(I)
C         RETURN
C         END
C
C         SUBROUTINE ANMODE
C         RETURN
C         END
C
C         SUBROUTINE MOVABS(I,J)
C         RETURN
C         END
C
C         SUBROUTINE DRWABS(I,J)
C         RETURN
C         END
```

APPENDIX II

LISTING OF
THE SHMO PROGRAM

```
C      PROGRAM EIGENVALUE
C BY G. PETTIT, THE UNIVERSITY OF MINNESOTA.
C
C      MODIFIED FOR USE OF THE OMEGA-TECHNIQUE WITH HUCKEL M.O.
C CALCULATIONS BY D. LAZDINS.
C
C      TRANSLATED FROM FORTRAN TO SCATRAN BY EVAN L. BRILL.
C FURTHER MODIFIED BY DOUGLAS FLECKNER, TO HANDLE ATOMS OF VARYING CORE
C CHARGE. BOTH OF THE OHIO STATE UNIVERSITY NUMERICAL COMPUTATION
C LABORATORY.
C
C      SUBROUTINES REORD AND BDORFV, AND FORTRAN IV TRANSLATION BY
C J.W. COOPER, OHIO STATE UNIVERSITY CHEMISTRY DEPARTMENT, 6/68.
C
C
C      THIS PROGRAM DETERMINES THE EIGENVALUES OF A REAL SYMMETRIC MATRIX
C A. ONLY THE TRIANGULAR SECTION TO THE RIGHT OF THE DIAGONAL,
C INCLUDING THE DIAGONAL  NEED BE ENTERED.
C      THE USUAL HUCKEL MATRIX CONSISTS OF DIAGONAL ELEMENTS (ALPHA - E),
C AND OFF DIAGONAL ELEMENTS BETA AND ZERO.
C
C      ALPHA AND BETA ARE VARIED BY THE FOLLOWING EQUATIONS.
C
C          ALPHA(X) = ALPHA(C) + H(X) * BETA
C
C          BETA(X-Y) = K(X-Y) * BETA(C-C)
C
C      WHERE C REPRESENTS CARBON, AND X AND Y ARE HETEROATOMS.
C
C      ONLY THE VALUES OF H AND K ARE ACTUALLY ENTERED ON CARDS.
C ALL OTHER POSITIONS IN THE MATRIX ARE 0.0)
C      TO VARY ALPHA, ENTER THE DESIRED VALUE OF H(X) IN THE DIAGONAL
C ELEMENTS. TO VARY BETA, ENTER K(X-Y) FOR EACH BETA. IF ONLY THE
C USUAL ALPHA AND BETA FOR CARBON ARE USED, THE DIAGONAL ELEMENTS
C ARE SET = 0.0, AND BETAS = 1.0 .
C
C      A TABLE OF COEFFICIENTS H AND K HAS BEEEN COMPILED BY L.B. KIER,
C BATTELLE MEMORIAL INSTITUTE, COLUMBUS, OHIO.
C
C      N = NUMBER OF ATOMS (DIMENSION OF THE MATRIX)
C
C      W = VALUE OF OMEGA
C
C      ITER = NUMBER OF ITERATIONS
C
C      NE = NUMBER OF PI-ELECTRONS
C
C      EP = AMOUNT OF DIFFERENCE DESIRED BETWEEN PRECEDING CHARGE DENSITY
C           AVERAGE, AND LAST CHARGE DENSITY.
C
C      IF OMEGA IS TO BE VARIED, SET W = LARGEST VALUE OF OMEGA TO BE
C USED, VAL= SMALLEST VALUE TO BE USED, AND STEP = INCREMENT BETWEEN
C SUCCESSIVE VALUES.
```

356

```
C      THE PURPOSE OF ITER IS TO STOP THE CALCULATION IF THE PROBLEM
C DOES NOT CONVERGE. THE PROPER NUMBER OF ITERATIONS TO ACHIEVE
C CONVERGENCE IS FOUND BY TRIAL AND ERROR. CONVERGENCE OCCURS WHEN THE
C ATOMIC CHARGE DENSITIES IN THE LAST TWO ITERATIONS DIFFER BY EP.
C      THE FREE VALENCE CALCULATION IS APPLICABLE TO CARBON ONLY.
C F(MAX) IS SET = 4.732, AND THE BOND ORDERS SUBTRACTED FROM IT.
C CALCULATIONS INVOLVING HETEROATOMS WILL HAVE CORRECT BOND ORDERS,
C HOWEVER, AND USING ANY DESIRED F(MAX) THESE FREE VALENCES CAN
C BE CALCULATED MANUALLY.
C
C      THE MAXIMUM VALUE OF N IS 30.
C
C
      DIMENSION A(30,30),B(30,30),C(30),D(30),AA(30,30),BB(30,30),
     1 CC(30),DD(30,30),EE(30),NCOL(10),COMT(18),FNUM(30),SSUM(30),Q(30)
     2, LBUF(33)
      COMMON A,AA,D,DD,Q,SSUM,N,NM,ITER,IT,W,VAL,STEP,NEL,LPT,EP,COMT,NE
      W=0
      EP=0
      ITER=0
      NM=0
C***************
C DEFINE LPT UNIT HERE
C FOR DEC-10 THIS WOULD BE 3
C BUT WE SPOOL IT TO A FILE 'LPT.TMP AND ONLY PRINT IF DESIRED
      LPT=21
      OPEN(UNIT=21,FILE='LPT.TMP',DISPOSE='DELETE',ACCESS='SEQINOUT')
C******
C*    1 READ(5,5)(COMT(I),I=1,18)
    5 FORMAT(18A4)
C*      READ(5,2)N,IEGEN,W,VAL,STEP,ITER,NE,NEL,EP,MOWGLI,(FNUM(I),I=1,N)
C*    2 FORMAT(2I4,3F4.2,3I4,F7.5,I4/(18F4.2))
C GET NUMBER OF ATOMS, NUMBER OF PI-ELECTRONS AND SET OTHER PARAMATERS
      IEGEN=0
      MOWGLI=1
C TYPE NAME FOR STARTU
1       TYPE 100
100     FORMAT(///' SHMO'/' NAME OF MOLECULE: ',$)
        ACCEPT 5,COMT
        TYPE 102
102     FORMAT('+NO. OF ATOMS= ',$)
        ACCEPT 104, N
        IF((N.LE.1).OR.(N.GT.30))GO TO 900
C       GET # OF PI ELECTRONS AND SET NEL
        TYPE 103
103     FORMAT('+NO. OF PI-ELECTRONS= ',$)
        ACCEPT 104, NE
104     FORMAT(I)
        NEL=N-2*(N/2)
      WRITE(LPT,6)(COMT(I),I=1,18)
    6 FORMAT(1H1,18A4)
C GET CORE CHARGES OF MOLECULE, USUALLY THSE ARE 1.0
C FOR EACH ATOM, UNLESS IT IS ACTUALLY CHARGED
        TYPE 105
105     FORMAT('+CORE CHARGES: ',$)
        ACCEPT 106,(FNUM(I),I=1,N)
106     FORMAT(20F)
C NOW SEE IF WE WILLBE DOING OMEGA TECHNIQUE
        TYPE 160
160     FORMAT('+OMEGA ITERATION?(Y OR N): '$)
        ACCEPT 121, ANS
        IF(ANS.NE.'Y')GO TO 165
C SET OMEGA TO 0 ORIGINALLY
        W=0
        EP=0
        ITER=10
        TYPE 161
161     FORMAT('+W=1.40, EP=.001 (Y OR N): '$)
        ACCEPT 121, ANS
        VAL=1.4
        W=1.4
        EP=.001
        IF(ANS .EQ. 'Y')GO TO 165
```

```
          TYPE 162
162       FORMAT(' OMEGA ='$)
          ACCEPT 106, W
          TYPE 163
163       FORMAT('+EPSILON=0'$)
          ACCEPT 102, EP
165         WRITE(LPT,7)EP,W,(I,FNUM(I),I=1,N)
    7 FORMAT(1H0,4HEP =,F8.5,10X,3HW =,F8.5//,10X,12HCORE CHARGES//12(I5
    1,F5.2))
    8 NM=NM+1
    9 WRITE(LPT,10)NM
   10 FORMAT(14H0 INPUT MATRIX,I5,2X,8HBY ROWS.)
C GET THE INPUT MATRIX HERE FROM THE TTY
          TYPE 107
107       FORMAT(' ENTER UPPER DIAGONAL OF MATRIX'/)
          DO 13 I=1,N
C*        READ(5,11)(A(I,J),J=I,N)
C*   11 FORMAT(9F8.3)
          TYPE 108, I
108       FORMAT(1H+,I3,2X,$)
          ACCEPT 109, (A(I,J),J=I,N)
109       FORMAT(20F)
       DO 12 J=1,N
   12 A(J,I)=A(I,J)
   13 WRITE(LPT,14)I,(A(I,J),J=1,N)
   14 FORMAT(/I3,F9.3,9F12.3/(10F12.3))
       DO 15 I=1,N
       DO 15 J=1,N
   15 AA(I,J)=A(I,J)
       DO 16 I=1,N
   16 D(I)=0.0
       IT=1
       DO 17 I=1,N
       SSUM(I)=0.
       DO 17 J=1,N
   17 SSUM(I)=ABS(A(I,J))+SSUM(I)
   18 CALL HDIAG(A,N,IEGEN,B,NR)
19        CALL REORD(B,CC)
20          WRITE(LPT,6)(COMT(I),I=1,12)
          WRITE(LPT,35)NM,IT,W
   35 FORMAT(1H0,21HEIGENVALUES OF MATRIX,I3,5X,13HITERATION NO.,I4,
      110X,4HW = ,F5.2/)
C PRINT LABEL FOR ENERGY LEVELS ON TELEPRINTER
C DETERMINE HOW MUCH OF TITLE BUFFER TO TYPE OUT
          DO 200 I=18,1,-1
          IF(COMT(I).NE.'   ')GO TO 201
200       CONTINUE
201       IX=I
          TYPE 110,(COMT(I),I=1,IX)
110       FORMAT('0ENERGY LEVELS FOR '18A4)
          WRITE(LPT,21)(I,CC(I),I=1,N)
   21 FORMAT(10(I3,F9.5))
C PRINT OUT LEVELS, ONE PER LINE ON TTY
          DO 111 I=N,1,-1
C TYPE OUT WITH - SIGN IF NEGATIVE TO PREVENT EXTRA + SIGNS
          CI=-CC(I)
          IF(CC(I).GE.0) TYPE 112,CC(I)
          IF(CC(I).LT.0)TYPE 113,CI
112       FORMAT(' ALPHA +',F7.3,' BETA  -----')
113       FORMAT(' ALPHA -',F7.3,' BETA  -----')
111       CONTINUE
       ENERGY=0.
       NE2=NE/2
       DO 235 I=1,NE2
  235 ENERGY=ENERGY+2.*CC(I)
       IF(NEL.EQ.1)ENERGY=ENERGY+CC(I+1)
       WRITE(LPT,335) NE,ENERGY
C TYPE OUT TOAL ENERGY AS WELL
          TYPE 335, NE,ENERGY
  335 FORMAT(1H0,10X,14HTOTAL ENERGY =,I4,8H ALPHA +,F6.2,5H BETA )
       IF(IEGEN.NE.0)GO TO 1
   22 WRITE(LPT,23)NM
   23 FORMAT(//23H EIGENVECTORS OF MATRIX,I3)
       JAY=0
```

```
   24 JOKER=0
   25 NV=0
   26 IF(N- (NV+10))27,28,29
   27 MAX=N-NV
      JOKER=1
      GO TO 30
   28 JOKER=1
   29 MAX=10
   30 DO 31 I=1,MAX
   31 NCOL(I)=NV+I
      INIT=NCOL(1)
      LAST=NCOL(MAX)
      WRITE(LPT,34)(J,J=INIT,LAST)
   34 FORMAT(1H0,10(I6,6X))
      DO 334 I=1,N
  334 WRITE(LPT,234) I,(DD(I,J),J=INIT,LAST)
  234 FORMAT(/I3,F9.5,9F12.5)
      IF(JOKER)37,37,38
   37 NV=NV+10
      GO TO 26
   38 IF(JAY)39,39,42
   39 DO 40 I=1,N
      DO 40 J=1,N
   40 BB(I,J)=DD(I,J)*DD(I,J)
      CONTINUE
   42 IF(NEL)1,43,44
   43 J=NE/2
      GO TO 45
   44 J=(NE-1)/2
   45 DO 49 I=1,N
      SUM=0.
      DO 46 M=1,J
   46 SUM=SUM+BB(I,M)
      IF(NEL)48,47,48
   47 Q(I)=FNUM(I)-SUM*2.
      GO TO 49
   48 Q(I)=FNUM(I)-SUM*2.-BB(I,J)
   49 CONTINUE
         IF(W.EQ.0)GO TO 250
      DO 249 I=1,N
  249 EE(I)=D(I)/W
  250     WRITE(LPT,50)(I,EE(I),I=1,N)
   50 FORMAT(//35H PREVIOUS AVERAGED ELECTRON DENSITY/(/10(I3,F9.5)))
      WRITE(LPT,51)(I,Q(I),I=1,N)
   51 FORMAT(//17H ELECTRON DENSITY/(10(I3,F9.5)))
C TYPE OUT ELECTRON DENSITY
         TYPE 140
  140     FORMAT('0ELECTRON DENSITY')
         DO 142 I=1,N
         TYPE 141,I,Q(I)
  141     FORMAT(I3,F6.2)
  142     CONTINUE
      IF(MOWGLI.EQ.1) CALL BDORFV
      IF(IT.EQ.ITER)GO TO 52
      IF(W.EQ.0.) GO TO 1
 2250     CALL OMEGA(NEXIT)
 2251     IF(NEXIT.EQ.1)GO TO 1
      GO TO 18
   52 WRITE(LPT,53) IT
   53 FORMAT(17H0****EXITED AFTER,I4,33HITERATIONS WITHOUT CONVERGING***
     1*)
      GO TO 1
C EXIT HERE AND PRINT OUT LPT OUTPUT ONLY IF DESIRED
  900     TYPE 120
  120     FORMAT(' PRINT OUTPUT ON LINE PRINTER?(Y OR N): '$)
      ACCEPT 121,ANS
  121     FORMAT(A1)
      IF(ANS.NE.'Y')STOP
C OTHERWISE REWIND OUTPUT FILE AND PRINT IT
      REWIND 21
  125     READ(21,126,END=130)LBUF
      WRITE(3,126)LBUF
  126     FORMAT(33A4)
      GO TO 125
```

```
130     STOP
        END




        SUBROUTINE ' HDIAG (H,N,IEGEN,U,NR)
C       MIHDI3, FORTRAN IV DIAGONALIZATION OF A REAL SYMMETRIC MATRIX BY
C       THE JACOBI METHOD.
C       CALLING SEQUENCE FOR DIAGONALIZATION
C       CALL HDIAG( H, N, IEGEN, U, NR)
C       WHERE H IS THE ARRAY TO BE DIAGONALIZED.
C       N IS THE ORDER OF THE MATRIX, H.
C       IEGEN MUST BE SET UNEQUAL TO ZERO IF ONLY EIGENVALUES ARE TO BE
C       COMPUTED.
C       IEGEN MUST BE SET EQUAL TO ZERO IF EIGENVALUES AND EIGENVECTORS
C       ARE TO BE COMPUTED.
C       U IS THE UNITARY MATRIX USED FOR FORMATION OF THE EIGENVECTORS.
C       NR IS THE NUMBER OF ROTATIONS.
C       A DIMENSION STATEMENT MUST BE INSERTED IN THE SUBROUTINE.
C       DIMENSION H(N,N), U(N,N), X(N), IQ(N)
C       COMPUTER MUST OPERATE IN FLOATING TRAP MODE
C       THE SUBROUTINE OPERATES ONLY ON THE ELEMENTS OF H THAT ARE TO THE
C       RIGHT OF THE MAIN DIAGONAL. THUS, ONLY A TRIANGULAR
C       SECTION NEED BE STORED IN THE ARRAY H.
        DIMENSION H(30,30),U(30,30),X(30),IQ(30)
        IF (IEGEN) 15,10,15
     10 DO 14 I=1,N
        DO 14 J=1,N
        IF(I-J)12,11,12
     11 U(I,J)=1.0
        GO TO 14
     12 U(I,J)=0.
     14 CONTINUE
     15 NR = 0
        IF (N-1) 1000,1000,17
C       SCAN FOR LARGEST OFF DIAGONAL ELEMENT IN EACH ROW
C       X(I) CONTAINS LARGEST ELEMENT IN  ITH ROW
C       IQ(I) HOLDS SECOND SUBSCRIPT DEFINING POSITION OF ELEMENT
     17 NMI1=N-1
        DO 30 I=1,NMI1
        X(I) = 0.
        IPL1=I+1
        DO 30 J=IPL1,N
        IF(X(I)-ABS(H(I,J))) 20,20,30
     20 X(I)=ABS(H(I,J))
        IQ(I)=J
     30 CONTINUE
C       SET INDICATOR FOR SHUT-OFF.RAP=2**-27,NR=NO.OF ROTATIONS
        RAP=7.450580596E-9
        HDTEST=1.E35
C       FIND MAXIMUM OF X(I) S FOR PIVOT ELEMENT AND
C       TEST FOR END OF PROBLEM
     40 DO 70   I=1,NMI1
        IF(I-1) 60,60,45
     45 IF(XMAX-X(I)) 60,70,70
     60 XMAX=X(I)
        IPIV=I
        JPIV=IQ(I)
     70 CONTINUE
C       IS MAX. X(I) EQUAL TO ZERO, IF LESS THAN HDTEST,REVISE HDTEST
        IF (XMAX) 1000,1000,80
     80 IF( HDTEST) 90,90,85
     85 IF (XMAX - HDTEST) 90,90,148
     90 HDIMIN =ABS(H(1,1))
        DO 110   I=2,N
        IF (HDIMIN - ABS(H(I,I))) 110,110,100
    100 HDIMIN=ABS(H(I,I))
    110 CONTINUE
        HDTEST = HDIMIN*RAP
C       RETURN IF MAX.H(I,J)LESS THAN(2**-27)ABSF(H(K,K)-MIN)
        IF (HDTEST-XMAX) 148,1000,1000
    148 NR= NR+1
```

```
C      COMPUTE TANGENT, SINE AND COSINE,H(I,I),H(J,J)
  150 TANG=SIGN (2.0,(H(IPIV,IPIV)-H(JPIV,JPIV)))*H(IPIV,JPIV)/(ABS (H(I
     1PIV,IPIV)-H(JPIV,JPIV))+SQRT ((H(IPIV,IPIV)-H(JPIV,JPIV))**2+4.0*H
     2(IPIV,JPIV)**2))
      COSINE=1.0/SQRT (1.0+TANG**2)
      SINE=TANG*COSINE
      HII=H(IPIV,IPIV)
      H(IPIV,IPIV)=COSINE**2*(HII+TANG*(2.*H(IPIV,JPIV)+TANG*H(JPIV,JPIV
     1)))
      H(JPIV,JPIV)=COSINE**2*(H(JPIV,JPIV)-TANG*(2.*H(IPIV,JPIV)-TANG*H
     1II))
      H(IPIV,JPIV)=0.
C      PSEUDO RANK THE EIGENVALUES
C      ADJUST SINE AND COS FOR COMPUTATION OF H(IK) AND U(IK)
      IF ( H(IPIV,IPIV) - H(JPIV,JPIV)) 152,153,153
  152 HTEMP = H(IPIV,IPIV)
      H(IPIV,IPIV) = H(JPIV,JPIV)
      H(JPIV,JPIV) =HTEMP
C      RECOMPUTE SINE AND COS
      HTEMP = SIGN  (1.0, -SINE) * COSINE
      COSINE =ABS  (SINE)
      SINE =HTEMP
  153 CONTINUE
C      INSPECT THE IQS BETWEEN I+1 AND N-1 TO DETERMINE
C      WHETHER A NEW MAXIUM VALUE SHOULD BE COMPUTE    SINCE
C      THE PRESENT MAXIMUM IS IN THE I OR J ROW.
      DO 350 I=1,NMI1
      IF(I-IPIV)210,350,200
  200 IF (I-JPIV) 210,350,210
  210 IF(IQ(I)-IPIV) 230,240,230
  230 IF(IQ(I)-JPIV) 350,240,350
  240 K=IQ(I)
  250 HTEMP=H(I,K)
      H(I,K)=0.
      IPL1=I+1
      X(I) =0.
C      SEARCH IN DEPLETED ROW FOR NEW MAXIMUM
      DO 320 J=IPL1,N
      IF ( X(I) -ABS ( H(I,J)) ) 300,300,320
  300 X(I) = ABS (H(I,J))
      IQ(I)=J
  320 CONTINUE
      H(I,K)=HTEMP
  350 CONTINUE
      X(IPIV) =0.
      X(JPIV) =0.
C      CHANGE THE ORDER ELEMENTS OF H
      DO 530 I=1,N
      IF (I-IPIV) 370,530,420
  370 HTEMP = H(I,IPIV)
      H(I,IPIV)= COSINE*HTEMP + SINE*H(I,JPIV)
      IF ( X(I) - ABS ( H(I,IPIV)) )380,390,390
  380 X(I) = ABS (H(I,IPIV))
      IQ(I) = IPIV
  390 H(I,JPIV) = - SINE*HTEMP + COSINE*H(I,JPIV)
      IF ( X(I) - ABS  ( H(I,JPIV)) ) 400,530,530
  400 X(I) = ABS (H(I,JPIV))
      IQ(I) = JPIV
      GO TO 530
  420 IF(I-JPIV) 430,530,480
  430 HTEMP = H(IPIV,I)
      H(IPIV,I) = COSINE*HTEMP + SINE*H(I,JPIV)
      IF ( X(IPIV) - ABS (H(IPIV,I)) ) 440,450,450
  440 X(IPIV) = ABS (H(IPIV,I))
      IQ(IPIV) = I
  450 H(I,JPIV) = - SINE*HTEMP + COSINE*H(I,JPIV)
      IF (X(I) - ABS ( H(I,JPIV)) ) 400,530,530
  480 HTEMP = H(IPIV,I)
      H(IPIV,I) = COSINE*HTEMP + SINE*H(JPIV,I)
      IF ( X(IPIV) - ABS ( H(IPIV,I)) ) 490,500,500
  490 X(IPIV) = ABS (H(IPIV,I))
      IQ(IPIV) = I
  500 H(JPIV,I) = - SINE*HTEMP + COSINE*H(JPIV,I)
      IF ( X(JPIV) - ABS ( H(JPIV,I)) )510,530,530
```

```
   510 X(JPIV) = ABS (H(JPIV,I))
       IQ(JPIV) = I
   530 CONTINUE
C      TEST FOR COMPUTATION OF EIGENVECTORS
       IF(IEGEN) 40,540,40
   540 DO 550 I=1,N
       HTEMP=U(I,IPIV)
       U(I,IPIV)=COSINE*HTEMP+SINE*U(I,JPIV)
   550 U(I,JPIV)= -SINE*HTEMP+COSINE*U(I,JPIV)
       GO TO 40
  1000 RETURN
       END




       SUBROUTINE OMEGA(NEXIT)
       DIMENSION A(30,30),AA(30,30),D(30),DD(30,30),Q(30),SSUM(30),C(30)
       COMMON A,AA,D,DD,Q,SSUM,N,NM,ITER,IT,W,VAL,STEP,NEL,LPT,EP,COMT,NE
C SUBROUTINE OMEGA VARIES ALPHA SO THAT
C      ALPHA(REAL) = ALPHA(C) + (1-Q(CALC)) * W * BETA(C-C)
       NEXIT=0
    52 I=1
    53 IF(ABS(D(I)/W-Q(I))-EP)54,54,56
    54 IF(N-I)55,265,55
   265 WRITE(LPT,266)IT
   266 FORMAT(1H0,30H****CONVERGENCE ACHIEVED AFTER,I4,14H ITERATIONS***)
       GO TO 65
    55 I=I+1
       GO TO 53
    56 IF(ITER-IT)1,58,57
    57 IT=IT+1
       IF(IT-2)58,60,58
    58 DO 59 I=1,N
    59 D(I)=(D(I)+W*Q(I))/2.
       GO TO 62
    60 DO 61 I=1,N
       D(I)=W*Q(I)
    61 C(I)=Q(I)
    62 DO 63 I=1,N
       DO 63 J=1,N
    63 A(I,J)=AA(I,J)
       DO 64 I=1,N
    64 A(I,I)=A(I,I)+D(I)
     1 RETURN
C
C      THE VALUE OF W IS VARIES FROM W TO VAL IN INCREMENTS OF STEP
C
    65 IF(W-VAL)68,68,66
    68 NEXIT=1
       RETURN
    66 W=W-STEP
       IT=2
       DO 67 I=1,N
    67 D(I)=W*C(I)
       GO TO 60
       END




       SUBROUTINE REORD(B,CC)
C THIS ROUTINE SORTS THE EIGENVALUES INTO DECREASING ALGEBRAIC ORDER.
       DIMENSION A(30,30),AA(30,30),D(30),DD(30,30),Q(30),SSUM(30),
      1COMT(18),CC(30),B(30,30)
       COMMON A,AA,D,DD,Q,SSUM,N,NM,ITER,IT,W,VAL,STEP,NEL,LPT,EP,COMT,NE
       Z=A(1,1)
       L=1
       DO 4 M=1,N
       DO 3 I=1,N
       IF(A(I,I).LE.Z) GO TO 3
     2 Z=A(I,I)
       L=I
```

```
    3 CONTINUE
      A(L,L)=-9999.9
      CC(M)=Z
      Z=-9999.9
      DO 4 J=1,N
    4 DD(J,M)=B(J,L)
      RETURN
      END




      SUBROUTINE BDORFV
C
C SUBROUTINE BDORFV CALCULATES THE BOND ORDERS AND FREE VALENCES
C     BY J.W. COOPER 1/68
      DIMENSION A(30,30),AA(30,30),D(30),DD(30,30),Q(30),SSUM(30)   ,
     1BSUM(30),F(30),P(30,30),COMT(18)
      COMMON A,AA,D,DD,Q,SSUM,N,NM,ITER,IT,W,VAL,STEP,NEL,LPT,EP,COMT,NE
      WRITE(LPT,6)(COMT(I),I=1,18)
    6 FORMAT(1H1,18A4)
C TYPE OUT BONDORDER MESSAGE
      TYPE 100
  100 FORMAT('0BOND ORDERS')
C TYPE OUT HEADER NUMBERS
      TYPE 110,(I,I=1,N-1)
  110 FORMAT(5X,20(I2,4X))
  268 WRITE(LPT,69)NM,IT
   69 FORMAT(1H0,12HBOND ORDERS,10X,10HMATRIX NO.,I3,5X,13HITERATION NO.
     1,I3//)
      KK=NE/2
      DO 71 J=1,N
      DO 71 I=1,N
      P(I,J)=0.0
      DO 70 M=1,KK
      IF(I.EQ.J) GO TO 70
      P(I,J)=2.*DD(I,M)*DD(J,M)+P(I,J)
   70 CONTINUE
      IF(NEL)80,71,270
  270 MM=KK+1
      P(I,J)=DD(I,MM)*DD(J,MM)+P(I,J)
   71 CONTINUE
      DO 271 I=1,N
      DO 271 J=1,N
      IF(ABS(AA(I,J)).LE.0.01)GO TO 271
      IF(I.EQ.J) GO TO 271
      P(I,J)=P(I,J)+1.0
  271 CONTINUE
      DO 273 I=1,N
      WRITE(LPT,72)(J,I,P(I,J),J=1,I)
C SKIP OVER LISTING FOR 1,1
      IF(I.EQ.1)GO TO 273
      TYPE 111,I,(P(I,J),J=1,I-1)
  111 FORMAT(I3,20F6.2)
   72 FORMAT(5(3H P(,I2,1H,,I2,3H) =,F8.5,5X) )
  273 CONTINUE
      TYPE 101
  101 FORMAT('0FREE VALENCES')
      WRITE(LPT,272)NM,IT
  272 FORMAT(1H0,14HFREE VALENCES,10X,10HMATRIX NO.,I3,5X,13HITERATION N
     10.,I4,//5X,11HATOM NUMBER,5X,4HF(I)/)
      DO 78 I=1,N
      IF(SSUM(I).GE.3.) GO TO 73
      IF(SSUM(I).GE.2.) GO TO 74
      IF(SSUM(I).GE.1.) GO TO 75
   73 BASE=4.732051
      GO TO 76
   74 BASE=3.732051
      GO TO 76
   75 BASE=2.732051
      GO TO 76
   76 BSUM(I)=0.0
```

```
      DO 77 K=1,N
      IF(ABS(AA(I,K)).LE.0.01)GO TO 77
      BSUM(I)=P(I,K)+BSUM(I)
   77 CONTINUE
      F(I)=BASE-BSUM(I)
      TYPE 102, I,F(I)
102      FORMAT(I3,F7.2)
   78 WRITE(LPT,79)I,F(I)
   79 FORMAT(9X,I2,8X,F8.5)
   80 RETURN
      END
```

INDEX

365